ET

Three Quests in Philosophy

Etienne Gilson was one of the most influential intellectuals and philosophers of the twentieth century. Some have credited him with expanding the spectrum of philosophical thought that had previously been limited by nineteenth-century analysts and positivists. Gilson devoted six decades to the study of the major philosophical figures of the Middle Ages. His interpretations of them are justly seen as new and insightful, and have exercised enormous influence on research in philosophy and on its presentation in the classroom. A "Gilson Society" has been active for years, and the Institut catholique in Paris has created a Gilson Chair in Metaphysics. A French publisher has announced a multi-volume publication of his complete works.

These seven previously unpublished lectures - Gilson termed them "Quests" - represent his mature thought on three key philosophical questions: the nature of philosophy, "species," and "matter." These are issues of perennial and pertinent interest to both philosophers and scientists. Gilson presents them here with his characteristic clarity, sense, and humour.

THE ETIENNE GILSON SERIES 31

ETIENNE GILSON

Three Quests in Philosophy

- The Education of a Philosopher
- In Quest of Species
- In Quest of Matter

Edited by
Armand Maurer

Foreword by James K. Farge

PONTIFICAL INSTITUTE OF MEDIAEVAL STUDIES

Acknowledgement

This book has been published with the help of a grant from the Canadian Federation for the Humanities and Social Sciences, through the Aid to Scholarly Publications Program, using funds provided by the Social Sciences and Humanities Research Council of Canada.

Library and Archives Canada Cataloguing in Publication

Gilson, Étienne, 1884–1978
Three quests in philosophy / Etienne Gilson ; edited by Armand Maurer.

(The Etienne Gilson series, ISSN 0708-319X ; 31)
The education of a philosopher first published in French under title: Réflexions sur l'éducation philosophique. Translated by James K. Farge.
Includes bibliographical references and index.
Complete contents: The education of a philosopher – In quest of species – In quest of matter.
ISBN 978-0-88844-731-9

1. Philosophy. 2. Species – Philosophy. 3. Matter – Philosophy. 4. Philosophy and science. 5. Gilson, Étienne, 1884–1978. 6. Philosophers – France – Biography. I. Farge, James K., 1938- II. Maurer, Armand A. (Armand Augustine), 1915–2008 III. Pontifical Institute of Mediaeval Studies IV. Title. V. Series.

B2430.G41M28 2008 194 C2008-905200-5

Pontifical Institute of Mediaeval Studies
59 Queen's Park Crescent East
Toronto, Ontario, Canada M5S 2C4

www.pims.ca

MANUFACTURED IN CANADA

Contents

Foreword

This slim volume contains seven previously unpublished lectures by Etienne Gilson.[1] He delivered the first of them, "The Education of a Philosopher," in Montréal in 1963. The next three, grouped under the title "In Quest of Species," were delivered in Toronto in January 1972. Gilson composed the last three, which he titled "In Quest of Matter," at his home in Cravant (Yonne), France; but his advanced age and declining health prevented his travelling to Canada. He therefore sent them to Laurence K. Shook in Toronto with the hope that they might eventually be published. "That is why I am anxious to do the job," wrote the late Father Armand Maurer, Gilson's student and disciple, in 2006.[2]

The first lecture was prompted when a group of students in philosophy at the Université de Montréal invited Gilson to speak at their inaugural "Semaine de Philosophie" on Tuesday, 19 March 1963. An audio tape recording was made, and some of the students later typed a transcription of it for private circulation. For forty-four years the lecture remained unknown except to those students who were present to hear it.[3] Then, in June 2007, Dr Raymond Fredette of Fitch Bay, Québec, who had been one of those students, made its existence known to me in my capacity as Librarian of the Pontifical Institute of Mediaeval Studies. He prepared a digital copy from the original typescript and sent it to me.

1. Just before going to press, I have learned that the French version of the first lecture has been published under the title, "Réflexions sur l'éducation philosophique," with no annotations, in the journal *Conférence* 26 (2008): 611–631. The following correction to its introductory note should be made: the French text came to us directly from Dr Raymond Fredette, not from the Institute archives. I gave a copy to Professor Brian Stock, who then communicated it to the editors of *Conférence*.
2. Letter to Father Lawrence Dewan, OP, in November 2006.
3. Laurence K. Shook, *Etienne Gilson* (Toronto, 1984), makes no mention of it.

It was with great excitement and pleasure that Armand Maurer first read the lecture by his esteemed friend reflecting on his own career as a philosopher and on the principles that guided him in preparing others for that career. Father Maurer clearly felt as if he were reliving some of the experiences he and Gilson had shared first as student and teacher and later in their long professional collaboration. Thinking that it might be published with the six other lectures by Gilson on which he had been working for some time, he asked me to translate it into English.

At the same time, however, Father Maurer was somewhat troubled by what seemed to be an inconsistency in the lecture which, he feared, might confuse some of its readers. In a short note to me about this, he pointed out how, not far into the lecture, when Gilson is speaking of the differences between philosophy and the sciences, he seems to limit the philosopher's field of interest to concepts (which he does not define) and to reserve the study of reality to the scientist. He added that, in this first instance, Gilson "sounds like a Platonist – something he surely was not!"[4] In the same note to me, however, Father Maurer cited the later section of the lecture where Gilson takes quite a different tack when he says, "philosophy is really interesting only when it

4. An anonymous reader of this manuscript has commented that, very early in his career, Gilson elaborated upon philosophy as concerned with the necessary, impersonal sequences of ideas. See "Le rôle de la philosophie dans l'histoire de la civilisation," in *Proceedings of the Sixth International Congress of Philosophy, Harvard University, ... 1926*, ed. Edgar Sheffield Brightman (New York, 1927), 529–535. Ten years later, however, Gilson was careful to note that the origin of those ideas, or concepts, is grounded in and governed by reality (*Le réalisme méthodique* [Paris, 1936], ch. 3). For translations of this latter source see the Bibliography of Works Cited.

Armand Maurer himself had commented on this subject in a note written at an earlier time and only recently found among his papers: "Gilson argued that youth is no time for metaphysics; yet he began with it, and only later in life did he delve into the philosophy of nature or physics. Aristotle started with physics and came later to metaphysics."

concentrates on reality." He urges the aspiring philosopher to "start with a real object and a real knowledge of that object," and then hammers home the point with this: "One can philosophize about everything provided that it be about *something*." Father Maurer was now confident that these remarks, coming at the end of the lecture, would reveal the real position of the Gilson he knew so well. He also sensed that the seven lectures published together could serve to throw light on the long, continuing debate about whether Gilson believed there is a philosophy of nature.[5] In his note to me, Father Maurer concluded, "In the [six lectures on species and matter] he accepts a philosophy of nature – and not just of the concept of nature but in the light of metaphysics." Finally, just two days before his death on 22 March 2008, he told me to publish the first lecture – just as he had originally planned – as an introductory essay to the other six lectures. He felt confident that readers would recognize in them Gilson's position on this controversy, adding that the six lectures on species and matter illustrate and confirm, each in its own distinct way, the principles which Gilson expounds in the first one, "The Education of a Philosopher." As was his custom, Armand Maurer was philosophizing to the very end.

Father Maurer had worked especially hard on editing the three lectures which Gilson entitled "In Quest of Species" and in composing the "Introduction" to them. The subject of these lectures lent itself in a special way to Armand Maurer's life-long love of science and his conviction that science and philosophy should work, each in its own sphere, to elucidate truth. But his waning strength and final illness prevented him from devoting equal time

5. On this see Ralph Nelson, "Two Masters, Two Perspectives: Maritain and Gilson on the Philosophy of Nature," in *Wisdom's Apprentice: Thomistic Essays in Honor of Lawrence Dewan, O.P.*, ed. Peter A Kwasniewski (Washington, D.C., 2007), 214–236. [Armand Maurer died before he could read or comment on this article.]

to editing the three lectures which Gilson called collectively "In Quest of Matter" and prevented him from composing an introduction for them. To fill this gap, I have drawn on the short historical context which Laurence K. Shook provided for them in his biography of Gilson.[6] This seemed an appropriate solution, since Father Shook had corresponded with Gilson about the lectures and later studied them in order to comment on them in the biography. As well, Shook was attuned to Gilson's thought, since he was the translator of the fifth edition of Gilson's *Le thomisme*.[7] Years later, he began the translation of its sixth and final edition,[8] which Maurer reviewed and completed after Shook's death.

In editing the lectures, Father Maurer was intent on respecting the integrity of Gilson's texts. He thus placed square brackets around his own footnotes in order to distinguish his editorial interventions from the footnote references that Gilson had supplied. He kept the latter intact, even in cases where more modern editions had become available, although in some cases he and I supplemented them with bibliographical details. Noting that Gilson quoted Darwin's *Origin of Species* from at least two different editions – the sixth edition (London, 1872; several reprints), which was the last in which Darwin introduced changes, and the *Great Books of the Western World* edition (Chicago, 1952) – we have cited both in each case. Unless otherwise noted, English translations in the texts of the six "Quest" lectures are by Gilson himself. The Bibliography of works cited (compiled by me) comprises works used by Gilson, Maurer, and me.

I am grateful to Father Lawrence Dewan, OP, who discussed several aspects of this manuscript with me, and to Prof. R. James Long, who supplied some references for it. The two anonymous

6. *Etienne Gilson*, 388–389.
7. *The Christian Philosophy of St Thomas Aquinas* (New York, 1956; repr. Notre Dame, Ind., 1983, 1994).
8. *Thomism: The Philosophy of Thomas Aquinas* (Toronto, 2002).

appraisers who read the manuscript for the Institute's Department of Publications made several helpful suggestions, especially about footnote references to the texts.

Etienne Gilson was seventy-nine years old when he spoke to the students in Montreal. He was eighty-eight when he delivered the lectures on species in Toronto and eighty-nine when he composed those on matter. In this thirtieth anniversary of Gilson's death, which occurred on 19 September 1978, the Pontifical Institute of Mediaeval Studies is pleased to publish these seven examples of his mature thought. We see them as a contribution to the renewed interest in Gilson's work and career that is manifested by the activities of the "Gilson Society for the Advancement of Christian Philosophy," by the creation, five years ago, of a Gilson Chair in Metaphysics at the Institut catholique in Paris, and by the growing spate of monographs, theses, articles, editions, and translations about and by Gilson. The recent announcement by his French publisher, Librairie Vrin, of its intention to publish the *Oeuvres complètes* of Gilson is only one more confirmation of the enduring value of his life's work.

James K. Farge

I

The Education of a Philosopher

Etienne Gilson

I received and accepted your invitation in a spirit of friendship, for that is the only appropriate response when young people do their elders the honour of making them a part of their concerns. It is in that same spirit of friendship that I have chosen the subject of this talk. It seems to me to be only natural that, in the course of – or, rather, at the end of – a long career and always seeking what is best for philosophy, a professor should pass on to students whatever his[1] personal experience of philosophy and that of his predecessors have taught him about what the education of a philosopher should comprise.

By philosophical education I have in mind here the whole span of the philosopher's life, from the time a young person first undertakes the study of philosophy. This is because, if that person's goal is to teach philosophy later in life, that process of teaching it will constantly produce new opportunities to learn about it. No one can ever permit himself to say about philosophy, "I know it." This concerns the very nature of what philosophy seeks to be. It requires vast knowledge, but it is not primarily a science; rather, it is a wisdom. No one can lay claim to be wise; one can only work hard to become wise. Such a life-long effort constantly to draw closer to wisdom is the very substance and reality of an education in philosophy. It is something the philosopher receives from others and something to which he can never cease to devote himself.

Are there signs of a vocation to become a philosopher? If so, what may they be? The first is a trait common to the philosopher and to all vocations which destine someone to a life of the spirit. Every person knows, acts, and produces or invents something. From that fact flows three principal kinds of life: speculation, action, or production. Although all of us partake of all three types, philosophy belongs primarily to the first. A strong desire to know,

1. [Translator's note: Gilson's use of masculine pronouns when referring both to men and women was standard grammatical usage in 1963. It has for the most part been maintained in this translation.]

to understand, to grasp uniquely or at least chiefly for the pleasure of understanding truth which is the good of the intellect: that is the first mark of a philosophical vocation.

Everybody desires to understand, but not all experience that desire to the extent of wanting to consecrate their lives to satisfy it. The future philosopher is one of those who do, and it is better that he realize it from the very beginning because, as a kind of vocation, he will experience some misgivings about it. In the course of his life he will see others who have devoted themselves to the disciplines of science, of action, or of production who have achieved magnificent careers, who perhaps have amassed fortunes, who accomplish good things for others in a more concrete way. If at such times the would-be philosopher tries to divide his efforts and join those others on their paths, that person is lost to philosophy. If you really want to become a philosopher, you will never be anything other than a philosopher. I can assure you this is a very good thing, but it is not everything. You will perhaps have some qualms about it, so you must be prepared for that.

A second sign of the philosopher's vocation is the particular nature of the knowledge that he wants to acquire. A student of Professor Martin Heidegger (I speak here of my friend Jean Wahl,[2] whom you all know) told me that in the opening lecture of a course, when Heidegger posed the classic question: "What is the object of philosophy? What is philosophy?" he naturally undertook to define its object. He told the class: "Take a look at the University's bulletin boards announcing the year's courses in the sciences, in the arts, and in philology, and take note of the varied subjects of those courses. The object of philosophy is everything which those subjects do not include." It is a fact that, even though the future philosopher can and ought to be interested in everything, the

2. [Like Gilson, Jean André Wahl (1888–1974) had also studied under Henri Bergson. He later lectured and wrote widely on Hegel and on Kierkegaard as well as on Heidegger.]

questions which interest him particularly are neither scientific nor literary, neither historical nor social. They are the questions that arise from no particular science but which the sciences presume to be resolved or which they resolve without perceiving in them any difficulty or even asking themselves questions about them.

The philosopher does not seek to understand the world – that is the business of the scientist – but he asks himself how is it that there is a world to understand? How is it that this world is intelligible to human beings and that there is an intelligent being to know it in its intelligibility?

Over and above every question of mathematics and physics – two sciences which still today bear the mark of his genius – the philosopher Leibniz asked himself, "Why is there *something* rather than *nothing*?" Now that is truly a question of philosophy. And, almost as if he wished to show that he was not only a philosopher but also a true philosopher, Leibniz quickly added this astonishing remark, "Because *nothing* is easier to understand than *something*."[3] Now that is the mark of a truly great philosopher! In its seeming naiveté, this question signifies that the complete absence of being would entail a complete absence of a question. If someone says to us, "There is *nothing* there; why is there *nothing* there?" there is no problem, there is no reason for *nothing* not to exist. But if someone tells us, "There is *something* there; why is there *something*?" then that is a real question. This is why *nothing* is easier to understand than *something*. I cite this story simply as an example of the mark of the true philosophical spirit. Now there we have an ancient question, one that Aristotle already named the "old question," a question ever posed, never resolved: *TI TO ON?* which means, "What does it mean to be?"[4]

3. [See "The Principles of Nature and of Grace, Based on Reason" 7, in Gottfried Wilhelm Leibniz, *Philosophical Papers and Letters*, ed. and trans. L.E. Loemker (Chicago, 1956), 2: 1036.]
4. [*Phys.* 1.1–3, 7, 9.]

Other questions of the same sort arise persistently in the same kind of spirit. "What do you want to know?" Saint Augustine asks himself; and he replies, "*Deum et animam scire cupio.* God and the soul, that is what I want to understand. Nothing else. Absolutely nothing else."[5] Those are still more questions and answers of philosophers.

One can recognize these questions using two signs: one positive, the other negative. First of all, these are questions that one ought to suppose to be resolved, even if one cannot resolve them, in order to be able to pose all sorts of other questions. That is the positive sign. If there is *nothing*, there is no question to be posed. But if there is *something*, then every question is posed. All the truly metaphysical questions are of that kind. That too is part of the positive sign. And now for the negative sign: these questions which present themselves to the spirit of every human being, if only ephemerally and without really pondering them, are precisely among those with which science cannot deal. It cannot do so, nor does it wish to do so, because the very possibility of doing science presumes them to be resolved. It is for that very reason that it declares them not only to be resolved but also to be useless and without object. This is because science presents reason in its pure application; and because, as far as science is concerned, whatever reason does not and cannot understand does not exist.

Anyone who commits to a philosophical career must therefore be someone for whom the importance of questions takes precedence in his mind over the demonstrability of answers. Aristotle said, and Thomas Aquinas often repeated after him, that the little we know about the most noble objects is worth more to us than the most perfect certitudes that touch on less perfect substances. The kinds of knowledge that we desire the most are those that give us the greatest joy.

5. [Augustine *Soliloquies* 1, 2, 7.]

Few scientific minds are so disposed. If your disposition is such that nothing can satisfy you except rational, experimental, clear-cut demonstration, then you are more cut out for the sciences than for philosophy. Science is a wonderful thing, it is indeed a beautiful thing; but don't launch out into a philosophical career if you have a mind that nothing can satisfy except a scientific demonstration. Only those who, with Aristotle, place a greater price on the importance of the object than on the certitude of demonstration are qualified for the study of philosophy. The examples of Hume and of Kant allow us to see what happens when a great mind that was made for the methods of science turns its hand to philosophy: it inevitably finishes by concluding that philosophical problems have no solutions simply because they have no mathematical or experimental solutions. It is useless to set out on a path that, although excellent in itself, can in no way take one where he wishes to go.

Let us suppose that our young person wants above all else to understand these basic and most noble of questions. That is an indispensable condition, but it is not yet sufficient to indicate a philosophical vocation. It is necessary as well that he have the kind of mind that suits the study of problems that a philosopher sets for himself. The basic matter of philosophical reflection is the concept. When someone asks you, "What is that?" and you respond "It's a horse," well and good. That is science. If that person adds, "What is the notion of horse?" well, that is philosophy. And if one enters into this kind of question – which a scientist has the good sense not to do, because as a scientist he can neither say nor do anything about it – then one is facing extremely complicated problems. The procedure that the future philosopher should expect to perform time and again consists in defining the meaning of the concepts that he uses. Regarding each concept he must explain each notion – that is the ideal that he ought to attain. He must explain, in relation to each of the notions that he uses, both what it is in itself and how it is distinct from every other concept.

And, as Plato once said, "For each occasion that something is itself, there are thousands of times that it is not something else."[6] One ventures out into very rough terrain.

The classic example of the way a philosopher thinks is the Socrates of the *Memorabilia* of Xenophon[7] and the Socratic dialogues of Plato – the Socrates whom Kierkegaard said was not a philosopher but was philosophy itself. In effect, Socrates, as Aristotle tells us, was concerned with concepts; but he said nothing about, and was in no way concerned with, nature. This does not mean that the philosopher can never speak also of nature; but when he does he is meddling in things he does not understand. It is the scientist, not the philosopher, whose business it is to speak about nature. When the philosopher does so, he acts as Saint Thomas did when he studied Aristotle or as we might do today in pretending to learn all about Einstein or about Louis de Broglie.[8] At such a time what we are really doing is stepping out of the realm of philosophy; we are leaving home base and entering into the field of science. In the contrary sense, the philosopher stays in his own territory when he concerns himself with concepts. The philosopher must constantly ask himself, "What am I at this moment trying to say?" First of all he is searching only to understand the meaning of the words he is using. That, by the way, is why he can quickly become insufferably boring, because no one can speak for long when he has to stop at each step of the

6. [The quotation has not been found; but Plato treats of this same matter in *Sophists* 256–257.]
7. [The *Memorabilia* of Xenophon (c. 428/7 B.C.–c. 354 B.C.) is also known as *The Memorable Things of Socrates*. Books 1 and 2 defend Socrates from the charges laid against him and deal with matters of education.]
8. [Louis de Broglie (1892–1987) was awarded the Nobel Prize for Physics in 1929 for his work with electrons, confirming experimentally (1927) his theory of particle-wave duality, i.e. that matter has the properties of both particles and waves.]

way to verify the sense of what he is saying. If your motive for becoming a philosopher is to ingratiate yourself with others, don't ever hope to succeed. One must never philosophize when he is invited to a friend's home, because it very quickly makes conversation impossible. Indeed, that is why, not knowing how to rid themselves of Socrates, the Athenians decided to poison him. When a philosopher enters into a conversation, it is usually he who empoisons everybody else. As some would put it, his being there makes it impossible to carry on a conversation. In common parlance, Socrates is one who splits hairs. In today's slang, the philosopher is a pain in the neck.

Not everyone has this somewhat troublesome vocation; but if one does have it, he must resign himself ahead of time to the inconveniences that go along with it. Compared to every other kind of writer, the philosopher will have a rather limited audience. But no one is ever obliged to be a philosopher.

There is even an evident contraindication, and this is the third point that I want to make. It is the opposite characteristic to the fault of those who barge into a conversation at any moment to ask what one is talking about. Let us name this quality "eloquence," the ability (as Cicero said) to speak well about everything and even about what one does not understand – to speak even better than those who do understand it. This is the masterful quality of the successful salesman or of the politician; it serves journalists and many others, too. But it is the scourge of philosophy. In this matter Cicero is the anti-philosopher *par excellence*.

There is such a thing as a philosophical style, and the future philosopher must work to acquire it. The good philosophical style is one which is barely visible, one which steps aside to let the idea, and it alone, be seen. One might say that it lets the thought emerge without any direct help from language. Ah! Now that is very much an art. But it is not one of the fine arts; it is art put at the service of truth, achieved with a great deal of renunciation. It is an art of which one must become a master.

Perhaps I can clarify what I'm trying to say in this comparison by posing the question, "Who is the perfect preacher, Bossuet or Bourdaloue?"[9] Everybody - well, everybody except preachers themselves - will respond "Bossuet." The preachers know full well that it is Bourdaloue, not Bossuet, who knows how to preach. In the same way, the philosopher who writes well, who has a beautiful philosophical style, is the one who doesn't even perceive that he has written, because he simply understands what he has tried to help the reader understand. Just as those who avidly seek empirically verifiable demonstrations would do well to dedicate themselves to the study of the sciences rather than to philosophy, so would those who are thrilled by elegance of style do well to put philosophy aside in order to give themselves over to the lively pleasure of literary composition and particularly to writing novels, which are the philosophy of the people.

A corollary which I want to formulate in this matter is addressed more to professors than to students. Professors of philosophy: take care not to consider a student unsuitable for philosophy just because he writes badly or because he has difficulty in expressing himself. If he writes badly, it is perhaps because he is trying to say something that is very difficult to say. If he has difficulties in expressing himself, it is perhaps because, contrary to normal conventions, he thinks before he speaks - or at least he is trying not to say anything that he does not think. Let us remember the "dumb ox"[10] of history. If the young Thomas Aquinas rarely

9. [Louis Bourdaloue (1632–1704) was a French Jesuit who could adapt his preaching style to audiences of every class and kind. The sermons of Jacques-Bénigne Bossuet (1627–1704) were considered to be models of the high French style. Voltaire, comparing Bourdaloue to Bossuet, preferred the former's simplicity and coherence.]

10. [Aquinas' earliest biographer, William of Tocco, recounts that Thomas' taciturnity and physical appearance drew this malapropos designation from some of his fellow students. See *Ystoria sancti Thome de Aquino de Guillaume de Tocco (1323)*, ed. Claire le Brun-Gouanvic (Toronto, 1996), ch. 12.]

spoke in class it was simply because he was waiting to have something to say. He saved what he had to say for later.

Let us suppose that our apprentice philosopher is endowed with all the correct aptitudes. How is he to proceed? There are several different ways to respond to this question, but one fact stands above all the others: it is that one does not philosophize alone. Philosophy is a conversation between philosophers. Someone who asks the ultimate questions proper to metaphysics, which is the highest form of philosophy, is not the first to do so nor will he be the last. Others before him have already asked themselves the same question, or else they ask them at the same time he does. And just like someone who wants to enter into a conversation already under way, our apprentice philosopher must know the context of the conversation; he must know what is being spoken about when he wishes to take his part in it.

There is no lack of contrary examples. We frequently see very learned men, for example those of scientific formation, suddenly decide to philosophize. But they are not the only ones: even self-taught men in that field (often, for reasons that I cannot fathom, engineers) take it on themselves to produce a "world system" in five or six volumes. Just this year I received the fifth volume of one of these. Those who do this are clearly men of great value and probably, in their own sphere, men of the first rank. Frequently their works are unreadable to the philosopher because, being self-taught, they are unaware of the current state of philosophical reflection, unaware of what problems have been put forth, unaware of what answers have already been proposed for them, so that they are forced to rediscover endlessly what we have long known or, to the contrary, to ignore important facts about the problems they are discussing.

Thus no one has any chance of inventing philosophy by himself, no more than a future scientist will have the chance of new discoveries without first knowing about ancient ones. In philosophy there is no such thing as a "self-made man." Like science, wisdom is learned at the feet of others. The only question is to know from whom to learn it. This time, I think, there are at least two valuable answers to this question.

As I recall from my own experience, and unless I'm deceiving myself, philosophical wisdom is for us first of all that of our own time. One might ask why I say "first of all," and I must admit that I do not know; I can only respond to this question, although it is perhaps not so much a question as a fact and an inevitable situation.

The disciples of Aristotle whom today we call Averroists and who believed in the indestructibility of species thought that, like the others, the species "philosophy" was indestructible. They thought that philosophy always exists somewhere. For them, when a philosopher dies in one part of the world, his philosophy dies with him; but, when another philosopher is born in another part of the world, philosophy is born again with him. It is perhaps a naive view of things, but at the same time it is a good example of the fact that there is almost always something true in what an intelligent man says. For, after all, there has certainly been a sort of circulation of philosophy in time and in space. Even though philosophy is an immaterial and eternal wisdom, it nevertheless has a history and even a geography. If we confine ourselves to what we more or less understand, that is, to the philosophy of the West, it seems to have been born in Asia Minor. From there, around the fourth century before Christ, it went to Greece and then, around the beginning of the Christian era, to Egypt. We then see it move to the Near East. From there it makes the tour of the Mediterranean, reappearing in Morocco, in Spain, from where it moves through the centuries to the North, as to a favourable land. We see it spreading today into new continents, where a new type of mind poses in its name new problems and proposes new solutions for the old ones.

Each one of us communicates first with philosophy under the form that it has had in a given time and place. That is why I would say that if someone asks himself where he ought to begin the study of philosophy, he has no choice in his answer. As a matter of fact, we begin with the philosophy that everybody is talking about. We begin by the philosophy that is talked about because we find it everywhere, in popular books and in newspapers, because it is the philosophy that everybody expects you to talk about. At the present time one cannot go anywhere to speak about philosophy without somebody saying to you, "Ah! Ah! You're going to speak to us about Heidegger." For the most part those people will not have read Heidegger, but they expect you to speak about Heidegger. They have in fact good reason for this: it is worth the effort. Or they will say to you, "Ah! You're going to speak to us about Teilhard de Chardin." Again, they may not have read Teilhard de Chardin. But those are the names that people talk about, these are the doctrines which are on everybody's lips, and they are the ones that appeal to young people from the moment they begin to think. I would not say that one must begin with that philosophy, but I do say that it is the philosophy by which one does begin.

When I was twenty years old, everybody in France began with Bergson.[11] Why? Well, it was very intelligent, because Bergson was by far the one person in France who was the most thoughtful in the matter of philosophy. After all, that was what was known, said, and spoken about, and young people found themselves caught up in a milieu that was saturated with those ideas and those movements. It was that philosophy which overpowered them, which chose them much more than they chose it. Any way you look at it, one cannot begin in any other way.

11. [Originally trained in mathematics, Henri Bergson (1859–1941) chose a career in philosophy instead. His lectures at the Collège de France drew large crowds; and his two major works, *Creative Evolution* and *The Two Sources of Morality and Religion*, exercised an extraordinary influence on French philosophy during his lifetime.]

But let us take a closer look. Some people judge this invasion of contemporary thought into the minds of young people inevitably to be a danger, the effects of which must be kept to a minimum. I don't say they are wrong, but I am not sure they are right. First of all, is it possible? How are you going to keep young people from listening to something that the whole world is talking about? But suppose it is possible. Is it desirable to protect young minds from the philosophy of their time? It goes without saying that they ought to be prepared so as to defend themselves against it, as well as to profit from what it has that is good. But that is not the crux of the question. In any event, it is important to acknowledge that the philosophy of our time is the only living philosophy, the only actually existing philosophy by which we can communicate with the philosophy that is eternal. The treasure of philosophical learning accumulated by wise men of all ages has a real existence only in the thinkers of today, in the mind of each one of us, in the present time in which we all take part.

In other words, there was a profound truth in that old doctrine of the perpetual reincarnation of philosophy that I spoke about a little while ago. Philosophy is ceaselessly reincarnated in new philosophers, because any philosophy that does not live again in this way is dead; it has ceased to exist. The oldest, the most classic of wisdoms is accessible to our thought and speech only in a language which is our own, which is actively occupied with resolving problems of our own time and of the state of the civilization in which we are destined to live.

Philosophical books are without any doubt very precious; but it is by means of the thought of our time, by our own thought, that we can communicate with the philosophy that is in the books. Philosophical education must therefore be generously and courageously open to the present. The future philosopher has no other means of preparing to take his place in the eternal circulation of wisdom than to resolutely take the plunge into the flow of present-day thought.

That is only half of the truth. What do we mean when we speak of the philosophy of our time? In this time that we call ours, philosophical production has seen such growth that the very idea of assimilating all of it has no meaning. Anybody who would naively attempt to do so would certainly go out of his mind. If we look at the extraordinary quantity of articles which appear in philosophical journals all over the world and in so many languages, how many are worth the trouble to read? It is much more difficult to say this than to answer the question: is it really necessary to begin by reading them all? This is materially impossible. When we receive four, five, six or seven philosophical journals each month, we don't even have the time to take them out of their wrapping. Instead we faithfully arrange them on our library shelves, where they will be conserved for many years.

As a matter of fact, we cannot expect to discover in any philosophical publication the ultimate wisdom – so long awaited – which will be the wisdom of tomorrow. Sometimes I am seized with a kind of vertigo when someone says to me, "Have you read such-and-such a book" by a recent philosopher? When I reply, "No," I'm asked, "But why not?" "Well, I haven't found the time; I had other things to read." "Yes, but this is the last word." At which point I tell myself, "This is pitiful; it's terrible; if we have had to wait until 1962 for the last word to have been said, or 1963 for the last word to be said in philosophy, then what is going to happen in 1964?" If wisdom has not yet been discovered (I'm speaking not of science, which is a completely different thing, because in fact the last word in science could quite plausibly be found on page 354 of a scientific journal that very few people other than rare specialists are capable of comprehending. That is something else; I speak here of philosophy), if philosophical wisdom has not yet been discovered, it never will be.

What we are looking to find in our time, because it exists nowhere else, is quite simply the philosophical life itself in which we seek to participate. It is almost certain that in the great pile of articles and books of philosophy published these days there is a large measure of hidden gold that we have come across too late to profit from before we in turn disappear. What is more, the philosophy that was alive when we were young threatens to keep us from perceiving the meaning of that which is alive today. But, just on that point, it happens that others have been, as it were, made by design to alert us to turn our eyes toward it and to help us penetrate its meaning.

In the past, one began philosophy with Aristotle; later, one did so with Descartes. Then it was Locke, then Husserl, and in my youth Bergson. Finally, today, I imagine that if I were beginning my philosophical apprenticeship, I could not escape the seduction of the disconcerting genius bearing the name Martin Heidegger. What I mean is that, in this immense river of philosophy, with its innumerable branches which in turn subdivide or end up in shallow basins that lead nowhere, there remains a strong central current where the great part of its water follows its course. And – what is so remarkable – everybody always knows where to find this central current. It could be that two or three principal branches open themselves to our choice, but so what? They finally come together again, and one or the other will lead us where we want to go, even though we can follow only one at a time. The choice is made for us, not by us. But it is made for excellent reasons which we will find impossible to understand unless, at the same time as in the philosophical present, we take the plunge courageously into its past.

This is why the study of the history of philosophy is necessary to the formation of a philosophical mind; and, as with the philosophy of our time, to study the philosophy of the past is a task that one must begin young but which is the work of a lifetime. Nobody can boastfully hope that, before he dies, he can read all the books of philosophy ever written. One must make a choice, and that choice will necessarily be tied to the nature of the very problems to which he seeks an answer.

But it is not by chance that certain names dominate the entire history of philosophy. Certain doctrines have expressed certain principal points of view on reality, and have done so in such a way that no one can philosophize without encountering those views under a form more or less resembling the one that those doctrines set up for them. Carlyle has written about the notion of representative men. It would be just as proper to speak of representative philosophies, each one of which is like the idea or the type of an ultimate truth about reality. I have already recalled the name of Socrates. One cannot, in fact, philosophize without defining the concepts which he uses. It is an eternal method, and anyone who has not the curiosity to be sure of the sense of the concepts he speaks about will never be a philosopher. Nor can one philosophize without looking to define what it is that makes reality intelligible, that is, without Platonizing. Einstein said, "It is strange that we are able to understand nature. What is completely incomprehensible is that understanding is possible." This is the very problem that Plato recognized: how is it that reality is intelligible? We have not advanced any further on this point than he did. The problem remains with us, and whoever wishes to look at it without reference to Plato simply does not know what he is talking about. Nor can one philosophize without seeking to know first principles and first causes; but that is to Aristotelize. And no one can push such research to its limit without ending with God who is the principle of principles and the cause of causes, and perhaps even beyond God into the One; and that is to Plotinize. One cannot not ask oneself if the ultimate questions of philosophy will find their answers in a wisdom which would not be philosophy but which would be higher still and which however would project from its high state its light on the problems that philosophy poses and on the intellect which seeks to resolve them; but that is to Thomistize. No one can philosophize without asking himself what are the relations between the truths of science and those of metaphysics; but that is to Kantize. Nor, finally, can one philosophize – and here I'm alluding to the very recent past, which is already past – without interrogating himself on the meaning of the modern doctrines of evolution and life and their relationship to ontology; but that is to Bergsonize.

Many other philosophers have said intelligent things, even profound things, and I have no intention of excluding those whom I have not named. But it would be hard to deny the fact that the effort to master the meaning of these lofty doctrines ought normally to be rewarded by an understanding that is more true and more profound, and by dominant philosophical truths of which their authors were the apostles and spokesmen in the world of philosophers.

This is certainly why the philosophy of our time remains incomprehensible if we isolate it from its past, of which it is only the prolongation and the continuation. It finds and renders to the past a present existence, fleeting but real, which is that of our own existence. The problems raised by each thinking link in the long chain of Western philosophy from Parmenides to our day come from farther away than we ourselves come, and we can comprehend them only in their reciprocal bonds. If we ask ourselves, "Where should the study of philosophy today begin?" there will always be someone there to tell us, "Begin with the philosophies of the past." What counts in the philosophical reflection that has developed is that which has been deployed over the course of the centuries, beginning with Parmenides of Elea and coming right down to our own day. Such is the life of wisdom, the wisdom of today which includes its own past. If it does not include its past it does not exist. It is thus good to initiate oneself into it early on and to grasp it, if I may put it this way, by both ends.

But it would be fatal to think that one can do this quickly. I said that no philosopher can be a "self-made man." I might add (and, judging from my experience with some wonderful North American students, this is not a useless observation) that there is also no way "to get rich quick" in philosophy. They want definitive and intelligible answers to questions that have occupied the finest minds since the beginning of western history – and they want them right away. This is not possible. There is no such thing as a young philosopher. There are only candidates to the dignity of philosophical maturity, and that only comes later.

Take careful note that this is something proper to philosophy. Mathematical geniuses generally make their discoveries at the age of twenty to twenty-five years – twenty-five being the outer limit. It takes longer for experimental scientists (who, as we have frequently seen, have careers as professors or politicians), and longer still for the ethical and political sciences. But it takes the longest to acquire metaphysics, which already in Antiquity was thought to be something one came to only after studying everything else.

I want to dwell somewhat on this, for many imagine that the meaning of that formula is more simple than it really is. They think it is merely a matter of planning in an orderly way their program of studies: first mathematics, then physics, then biology, followed by ethics, after which metaphysics comes at the end of the program. But neither Aristotle nor Saint Thomas would allow metaphysics to come, as they say in the schools, at the end of the program. Aristotle said, and he was only following Plato, that when it comes to the highest philosophy, those who are too young for it talk but do not think. Saint Thomas comments mercilessly on this passage that we ourselves repeat mercilessly to our young students when we are ourselves young professors. *Juvenes sapientialia quidem scilicet metaphysicalia non credunt, id est, non attingunt mente licet ea dicant ore.*[12] They speak about metaphysics. We who have all been young, we remember well. No one speaks so eloquently of metaphysics as when he is young. And the students of metaphysics, they speak admirably about it. That is why Plato said, "Don't let them talk, because they are then like young dogs amusing themselves with some game bird they have found. They

12. [*Sententia libri Ethicorum* 6.7.16.]

play around with it; and, being young, they will never really understand later what they were playing with."[13]

This is a very tough doctrine, as you can well see. And yet, Saint Thomas commented on it when he added that it was for that very reason that the ancients did not put this study at the end of their program of studies but that they reserved it for the end of their life.[14] It was not something for their final year of studies; it was purely and simply for the final years of their life.[15] Philosophy is not a profession in which the intellect grows rich quickly.

But this is no reason to become discouraged. On the contrary, as I have said, one must begin when one is young, and mostly because no one can ever prevent someone who wants to philosophize from doing so when he is young. That is the time of life when one has the mind for grand questions. But any young person who starts to philosophize must be warned that the goal is a long-term one. Anybody who thinks he can get there early will never get there. For such a one, the end of the voyage will always remain that moment when he thought he had arrived.

What does one do in the meantime? One must prepare oneself for the arrival of the light. And how does one prepare? By acquiring positive, scientific knowledge, whether of one or several orders of reality, precisely those of which one would like to know what are their first principles.

I said at the beginning of this talk that one can know he wants to be a philosopher if he asks himself the ultimate questions that

13. [Gilson is analyzing, rather than quoting, Plato, as the French transcript implies. See the *Republic* 7 539–540, in *Dialogues of Plato*, trans. B. Jowett, 1: 798–799, and Jowett's own "Analysis" of it: "Let us take every possible care that young persons do not study philosophy too early. For a young man is a sort of *puppy* who only plays with an argument and is reasoned into and out of his opinions every day; ... ."]

14. [Preface to *Super librum De causis expositio*, ed. H.D. Saffrey, OP (Fribourg; Louvain, 1954).]

15. [See above, p. 3.]

no science, as science, either wants to, or can, put to itself. I would like to add now that the future philosopher must ask those questions in relation to certain other less general ones, questions which interest him in themselves but which he knows do not themselves contain their ultimate response. It is in relation to one or the other of these disciplines that the future philosopher will pursue the search for first principles. One prepares oneself for the coming of his metaphysical age (I don't dare tell you the age at which Saint Thomas put it) with an initiation into, a waiting for, the technique of some science or particular discipline studied and practised for itself, but with the ever-present afterthought that the knowledge thus acquired must serve as a starting point towards acquiring the comprehension of the first principles of knowledge and thus also towards the acquisition of metaphysical wisdom. Such is the true sense of the teaching of Aristotle and of Saint Thomas on this critical point. Metaphysics is acquired last because it has meaning only in relation to an acquired positive knowledge of which it is the crowning achievement.

Let me insist on this point, because I believe the teaching of philosophy ranges over a broad spectrum. Goethe once wrote, "I have never thought about thought."[16] A philosopher could never allow himself that kind of detachment, because he must necessarily think about thought. One might, however, say ("with a grain of salt") about the philosopher that he should not try to philosophize about philosophy, and that doing so is what makes courses in philosophy so boring: they concentrate on philosophy, when philosophy is really interesting only when it concentrates on reality. One must never "metaphysicize" about metaphysics. What philosophy must do is lift itself above thinking about the sciences; it should turn its attention to what it knows, considering first prin-

16. [Johann Wolfgang von Goethe, *Zahme Xenien* 7.110. See *Sprüche in Reimen, Zahme Xenien, und Invektiven*, ed. Max Hecker (Leipzig, 1908). (Father Owen Lee provided this reference.)]

ciples in order to be sure of their meaning. It must therefore force itself, with no hesitation, to consider them in themselves. It must start over, again and again, beginning with experience of the senses, to make the same effort to connect them. The effort of lofty intellectual abstraction required for this is proper to the metaphysician. But he cannot remain at this level, and it is not his destiny to remain there. In fact, it is only at the end of an heroic effort to observe them at work in the interpretation of sensible reality that he will truly grasp the meaning of the principles he has attained. It is only in this sense that they are wisdom, insofar as they confer meaning and intelligibility to the knowledge of which they alone are the ultimate justification. One attains metaphysical understanding at the end of a long period of endeavour in some science, starting with something one knows and, by a habit acquired progressively, to interpret all knowledge in the light of first principles.

Saint Thomas said something on this subject that is well worth looking at – something that, as I recall, astonished me the first time I ran across it. As usual, he made it in regard to a very different problem. According to Saint Thomas one learns philosophy when dealing with something other than philosophy. He did it when faced with a theological problem: "What is the final end of man?" He naturally replied, "It is to know God."[17] He does not say that it is to acquire knowledge or even wisdom nor even the understanding of first principles. He says, "No, not that." The knowledge of first principles cannot bring happiness to man because it is not true knowledge: to know God is to know some one, that is, to apprehend a real object subsisting in itself and separate from the intellect which knows him. "This is why," he says in the *Summa contra Gentiles*, "the supreme happiness of man could never be the contemplation which consists in knowing principles. This, in effect, is very imperfect, because it is absolutely

17. [*Summa theologiae* 1-2.3.8c.]

universal and general, comprising the knowledge of things only in a virtual way."[18] If you are doing philosophy and think that you have attained the knowledge of principles, you are completely forgetting that such knowledge is abstract and that to know is to know objects in the light of principles, and that in reality one can never know the principles if he does not know the thing in the light of those principles.

Applying this remark to our problem, we can say that metaphysical wisdom, completely centred as it is on the search for principles, cannot be content to look at them in their state of abstraction. It consists rather in seeing them at work, caught up in concrete reality and rendered visible by the light which metaphysics throws on reality. The object of metaphysics is the knowledge of principles as light and cause of all reality.

It would therefore be an error to look for philosophical wisdom in the sole study of the first principles of understanding. The sterility of so many lives generously devoted to the study of wisdom comes from the fact that it is thought to be an object apart, knowable in itself, outside of the real that it illuminates and of the science of which it is the crown.

I would therefore pose this last question to anyone who wants to philosophize: "What do you want to philosophize about?" This can be anything you choose, provided that it be about some thing and not about philosophy. You can philosophize about physics, as Aristotle did, about ethics, as did the Stoics, about biology, as Bergson did. You can even perhaps risk philosophizing about the history of the philosophies of past times or about those of economic or social doctrines, or about the problems those sciences pose for thinking persons of our day. Start with any given fact in your experience of the present day and ask yourself: "What does metaphysical wisdom teach me about it? What does metaphysical

18. [*Summa contra Gentiles* 3.37.8.]

wisdom counsel me to think or not to think about these things and about the doctrines that pretend to interpret them?" The object itself is of little importance, as long as you start with a real object and a real knowledge of that object. By real knowledge I do not mean great erudition or very deep understanding but only that it be first-hand knowledge; and that the object, the intellectual object about which you speak, be an object that is alive in your own thought, something that you yourself have tried to elucidate. One can philosophize about everything provided that it be about something. Whenever philosophy tries to feed on its own substance it soon degenerates into an empty verbalism – intolerable to others and, in the end, discouraging to the one who practices it. If I may put this in very general terms, I would conclude with this: "Let your metaphysics always bear on the physical – 'physical' in the sense that Saint Thomas used it: 'sensate reality' – and see that it aims at being its highest elucidation."

My last word on philosophical education is therefore simply this: take courage and be patient in the search for the first truth and in the elucidation of reality in the light of that truth. This, as I said earlier, is a life-long task. Let me just add in conclusion that there is plenty of material for a life's work in philosophy and that it is well worth your while. It is not the supreme happiness of man, but it is its earthly image; and in this way it is joy: *gaudium de veritate*, joy flowing from truth.

II

In Quest of Species

Etienne Gilson

Introduction

Etienne Gilson composed the three lectures on species in English at his home in France in the summer and fall of 1971, and revised them twice. They therefore represent his well-considered and final views on the topic. The previous year, in January 1971, he had given three lectures in Toronto entitled "In Quest of Evolution," the contents of which will be found in his expanded treatment of the subject, *D'Aristote à Darwin et retour: Essai sur quelques constantes de la biophilosophie*.[1] While writing that book, Gilson became intrigued by the notion of species, which had puzzled even Darwin himself. He noted that Darwin wrote a good deal about the *origin* of species but said little about the *nature* of species; nor did the famous naturalist think that much could be said about it.[2]

While preparing his lectures on species, Gilson looked about to see what others, especially contemporary Thomists, were saying about the subject. Knowing that Mortimer J. Adler, a prominent Thomist at the University of Chicago, had written at length on the subject of species, Gilson requested that a copy of Adler's articles in *The Thomist* be sent to him in Paris. I do not know if he read all of them, but he read enough to realize that Adler's notion of species was diametrically opposed to his own. Adler was convinced that *species* is strictly a logical notion that as such is never used ontologically.[3] In adopting this position Adler was in agreement with Jacques Maritain, who wrote that "the notion of species

1. (Paris, 1971); translated by John Lyon as *From Aristotle to Darwin and Back Again: A Journey in Final Causality, Species, and Evolution* (Notre Dame, Ind., 1984).
2. Gilson touches upon Darwin's notion of species in Appendix II to his *D'Aristote à Darwin* ("Darwin en quête de l'espèce"), 225–248, and in the English translation ("Darwin in Search of Species"), 139–156.
3. Adler, "Solution of the Problem of Species," *The Thomist* 3 (1941): 297–298 n. 27.

is *in itself* a logical notion ... ."[4] To Gilson, however, although there is a logical notion of species, species are in the first place realities, and he confirms this by appealing, in three successive lectures, to common sense, science, and philosophy. Each of these in its own way has something to tell us about the reality of species, but philosophy is the most informative about their nature.

Gilson says that the main lesson we can learn from the experience of species is the certitude that they are real and not merely names or mental concepts devised for practical purposes or needs. He enforces the certitude of this empirical fact with both cogent arguments and examples, while conceding that it touches the fact of the reality of species rather than their nature. Incidentally, Gilson borrows the term "pure experience" from William James, not with all the Jamesian meaning it has in that philosopher's radical empiricism but simply as common sense or plain experience. James' account of pure experience is quite alien to Gilson's Thomistic realism and, according to him, less precise.[5]

Having established the reality of species through experience in Lecture 1, Gilson proceeds in the next lecture to consider what the scientists tell us about species. The naturalists whom Gilson consulted were mainly French thinkers from the eighteenth and nineteenth centuries: the naturalists Georges-Louis Leclerc, known as Comte de Buffon, Georges Cuvier, Jean-Baptiste Lamarck, and Lucien Cuénot. He also, of course, consulted Charles Darwin as well as the Russian geneticist Theodosius Dobzhansky. There has naturally been a wealth of later speculation about biological spe-

4. Jacques Maritain, foreword to Adler's *Problems for Thomists: The Problem of Species* (New York, 1940), 14. Other Thomists with the same notion of species could be cited, especially John N. Deely, "The Philosophical Dimensions of *The Origin of Species*," *The Thomist* 33 (1969): 75–149, 251–335.

5. See below, p. 37 n. 8. For Aquinas' doctrine of knowledge see Etienne Gilson, *Thomism: The Philosophy of Thomas Aquinas*, trans. Laurence K. Shook and Armand Maurer (Toronto, 2002), 241–273; Robert Pasnau, *Thomas Aquinas on Human Nature* (Cambridge, 2002), 171–90, 267–360.

cies that Gilson, in 1972, was not able to take into account. Some of it, however, confirms his view of the reality of species. Ernst Mayr, for example, writes, "Among theorists in biology a realist interpretation of the concept of species is most common, but the fundamentalist account also has some adherents."[6] Mayr criticizes "the nominalist conception that 'a person' (not nature) makes species by grouping individuals under a name ... . Nothing brought this point home to me more forcefully than the fact that the Stone Age primitive natives in the mountains of New Guinea discriminate and name exactly the same species that are distinguished by the naturalists of the West."[7] Michael Ruse is also a realist in his essay, "Biological Species: Natural Kinds, Individuals, or What?"[8]

Gilson emphasizes the fact that, according to Darwin, naturalists do not agree on the definition of a species, or even on that of the closely allied notion of a variety. Darwin himself was satisfied with a vague notion of a species.[9] Moreover, lacking a clear idea of what a species is, naturalists differ widely as to their num-

6. Cited by Bernd Graefrath, "Darwinism: Neither Biologistic nor Metaphysical," in *Darwinism and Philosophy*, ed. Vittorio Hösle and Christian Illies (Notre Dame, Ind., 2005), 373. See Philip Kitcher, "Species," in *The Units of Evolution: Essays on the Nature of Species*, ed. Marc Ereshefsky (Cambridge, Mass., 1992), 317–342.
7. Ernst Mayr, *This is Biology: The Science of the Living World* (Cambridge, Mass., 1997), 131.
8. *British Journal for the Philosophy of Science* 38 (1987): 225–242.
9. The closest Darwin comes to defining species is the following: "I look at the term species as one arbitrarily given, for the sake of convenience, to a set of individuals closely resembling each other, and that it does not essentially differ from the term variety ... [which] is also applied arbitrarily, for convenience's sake" (*The Origin of Species by Means of Natural Selection*, 6th ed. [London, 1872], 42; cf. Great Books of the Western World, ed. Robert M. Hutchins [Chicago, 1952], 49: 29). The editions will be referred to respectively as "6th ed." and "Great Books ed." hereafter. In his final chapter, Darwin writes of "the vain search for the undiscovered and undiscoverable essence of the term species" (Ibid. [6th ed., 426; Great Books ed., 242]).

ber. They are also at odds on whether only individuals are real, and whether species are nothing but products of our minds.

Naturalists are here involved in a problem that has long perplexed philosophers: what is the status of individuals and universals? Are only individuals real, or do universals enjoy some reality? Another possibility is that species alone are real and individuals lack stable existence. Until these questions can be answered, the quest for species is not finished.

Gilson takes up these questions in his final lecture about philosophy and species; but in an important digression at the end of Lecture 2, he discusses the part of philosophy in which their answers might be found. Many Thomists would say that they are to be found in the philosophy of nature, conceiving this as a major division of philosophy, formally distinct from mathematics, the sciences, and metaphysics. Although Gilson does not mention Jacques Maritain among these Thomists, no doubt he has him in mind.[10] Instead, he cites Mortimer J. Adler, among others, who followed Maritain's views on the subject.[11]

In none of his previously published works does Gilson clearly express his views on the philosophy of nature. He does not deny its existence, but neither does he wholeheartedly accept it in its usual Thomistic sense. He sometimes uses the term, but without much conviction. Some fault him for seeming to reduce all philosophy to metaphysics. To my knowledge, the only occasion on which he clearly spells out his views on the philosophy of nature is the present set of lectures on species. Here he denies that he has anything "against the project of a philosophy of nature"; the only question is the meaning of the term.[12] Neither Aristotle nor Thomas

10. See Jacques Maritain, *Philosophy of Nature*, trans. Imelda C. Byrne (New York, 1951). The original French version appeared in 1936.
11. Adler, *Problems for Thomists: the Problem of Species* (New York, 1940); Adler, "Solution of the Problem of Species," *The Thomist* 3 (1941): 297–298.
12. See below, pp. 55–56.

Aquinas saw any need for a philosophy of nature as a distinct branch of philosophy midway between physics and metaphysics. In physics, they treated of particular subjects like the four elements, animals, and humans, but they also rose above specific examples to consider the general aspects and principles of nature, such as matter and form, and these are the crowning part of the science. In this Aristotelian and Thomistic conception, Gilson did not think that the philosophy of nature should be considered as an autonomous branch of philosophy but as the crowning part of physics, just as (natural) theology is the crowning part of metaphysics.

It will be objected that this may have been true in medieval physics but not in modern physics, which has abandoned the Aristotelian notion of nature, particularly the notion of substantial form. Aristotle, and Aquinas after him, thought that there is something immaterial in nature, namely substantial form, which together with matter constitute a material thing. Since the time of Descartes, however, nature is generally reduced to matter and identified with "geometrical extension in space and the mechanical laws of motion." Modern scientists on the whole agree with Descartes and deny that there is anything non-material, like a substantial form, in a living being.

The result, according to Gilson, is that evident biological facts are unintelligible to the scientist. In the lectures on evolution in 1971 he pointed out that biologists cannot account for final causality, and so they deny its existence in living beings. In the present lectures he stresses that, although the biologist may accept the reality of species, he has difficulty in defining their nature, and the reason is the same: species, like final causality, is not strictly a scientific but a philosophical notion.

Gilson's third lecture on species is devoted to justifying this bold statement, which he does by recourse to the metaphysical notions of substantial form and being (*esse*). He has already shown that species are real for both common sense and science. There are Thomists, however, who contend that the concept of species is strictly logical and therefore exists only in the mind. How, then, can a species be real? Gilson's answer is that it is not real as abstracted from individual things and existing as a universal concept in the mind, where its existence is immaterial like that of the mind itself. But before the species exists as a logical concept it has a reality in things or substances, where it shares in their existence, specifying it as such-and-such – for example, a man.

Armand Maurer

1. *Species for Pure Experience*

Only late in life do some of us realize what confers upon our philosophical reflection some degree of unity. My first publication, as far as I remember, appeared in the *Revue philosophique* of 1909, under the title "On Absolute Positivism."[1] At that time absolute positivism was a new term, coined by Abel Rey, a professor of philosophy at the Sorbonne, in order to designate what we today more harshly call "scientism."[2]

I do not remember the words I used then, but I distinctly recall my state of mind in raising that youthful protest against scientism. I was naively but intensely feeling indignant about a university professor of philosophy brazenly teaching that there was no such thing as philosophy. The director of the *Revue philosophique* was my good master, the positivist and sociologist Lucien Lévy-Bruhl.[3] Incidentally this shows how vain are the scruples of some teachers who fear to exercise too strong an influence on the thought of their pupils. Lévy-Bruhl fundamentally agreed with Abel Rey, and I knew it well. Still, I gave him my critical article, and it is typical of his liberalism that he printed it, while letting me know that my own position was definitely out of date.

1. "Sur le positivisme absolu," *Revue philosophique de la France et de l'étranger* 68 (1909): 63–65.
2. [Gilson was replying to Abel Rey, "Vers le positivisme absolu," ibid., 67 (1909): 461–479. Armand Maurer reviewed this incident in Gilson's career in his article "Etienne Gilson, Critic of Positivism," *The Thomist* 71 (2007): 199–220. He developed some of his own ideas on the philosophical implications of evolution in "Darwin, Thomists, and Secondary Causality," *The Review of Metaphysics* 57 (2004): 491–514. JKF]
3. [Gilson wrote an obituary of his friend: "Mon ami Lévy-Bruhl, philosophe, sociologue, analyste des mentalités primitives," *Nouvelles littéraires*, 18 March 1939, 1. Gilson's debt to Lévy-Bruhl for his method of analyzing the thought of individual philosophers is stressed by Francesca Aran Murphy, *Art and Intellect in the Philosophy of Etienne Gilson* (Columbia, Mo., 2004), 33–35.]

If it was already out of date sixty-three years ago, it must be even more outdated today. But I am still of the same mind. More firmly than ever I believe in the specificity of the philosophical order and of its problems, which entails a corresponding belief in the specificity of the methods of philosophy and its conclusions.[4] I call "specifically philosophical" those problems that arise *in* science, *from* science, and for which there is no possible scientific conclusion.[5] In my lectures on final causality in 1970,[6] I had to stress the fact that many scientists mistakenly feel qualified to decide about the existence of final causality in nature, while purposiveness in living organisms is offered by nature to scientists as a fact, on the reality of which they are not even consulted.

While I was dealing with final causality, another problem was obstinately trying to force itself upon my attention, namely that of species. At the time, I noted the fact that in Darwin's *Origin of Species* the famous naturalist has dealt at great length with the *origin* of species but has said little about their nature. In this year's

4. [Gilson deals with the specificity of philosophy in ch. 3 of his *Réalisme méthodique* (Paris, 1936). An English translation of ch. 3 by D.A. Patton, "Concerning Christian Philosophy: The Distinctiveness of the Philosophic Order," appeared in *Philosophy and History: Essays Presented to Ernst Cassirer*, ed. Raymond Klibansky and H.J. Paton (Oxford, 1936), 61–76, and was reprinted as "The Distinctiveness of the Philosophic Order" in *A Gilson Reader*, ed. Anton C. Pegis (Garden City, N.Y., 1957), 49–63. For a new translation by Philip Trower of Gilson's entire book, see *Methodical Realism* (Front Royal, Va., 1990).]
5. [Gilson added this definition in the margin, probably when revising his text.]
6. [The four public lectures on "Finalism Revisited" were given at the Pontifical Institute of Mediaeval Studies, Toronto, in January, 1970. They are entitled: "The Case for Final Causality," "The Case for the Mechanical Cause," "Finalism and Physical Probability," and "Evolution: Teleology and Theology." For a description of them see Laurence K. Shook, *Etienne Gilson* (Toronto, 1984), 383–384, who sees them as the first of "the Scientific Trilogy, 1970–1972."]

lectures[7] I would like to take up that unfinished piece of business and to consider, in itself and for itself, the problem of the nature and reality of species. And my first remark, which might well be my last, is that, considered in itself, species is neither a scientific nor a philosophical notion. If there were no human beings, there would be neither philosophy nor mathematics; sciences of nature themselves would not exist; there would be no zoology; but there would still be species. Not, of course, the *notion* of species, which, like every notion, requires the existence of a knowing being, but the very reality which the notion of species signifies would undoubtedly exist. Had there been living species on the moon, the arrival of the astronauts would not have created them; they would have found them and collected some specimens of them, or at least photographed some of them for our information.

The actual existence of species in nature, quite apart from the fact that we do know them, is supported by the behaviour of dumb animals. They are dumb in the sense that they are without the power of speech, but not at all in the sense that they are stupid or without the power of knowledge. Animals do not use words. They have no verbal signs denoting abstract concepts, such as that of a given species, much less that of species in general; but they do know what we call certain species when they are vitally interested in knowing them. The possibility of animals' surviving implies their capacity to distinguish water from food, and fitting food from unfitting or harmful food. Animals can be very choosy in that respect. If they have the freedom of a garden, turtles unerringly make for strawberry beds; if there is a raspberry patch, they

7. [That is, January 1972. Gilson does not mention his three lectures "In Quest of Evolution" given on 13, 20, and 27 January 1971. The titles of these lectures were "Darwin without Evolution," "Evolution without Darwin," and "From Malthus to the Twilight of Evolutionism." See Shook, ibid., 385–386. The lectures on final causality and those on evolution are incorporated in his book, *From Aristotle to Darwin and Back Again* (see above, p. 27 n. 1).]

will prefer it; and if they have to fall back on grass and leaves, turtles will select dandelions, with a marked preference for the flower over the leaf.

One hesitates between the fear of uselessly expatiating on such immediate evidence and that of allowing it to pass unnoticed. What Goethe used to call *Wahlverwandschaften* (selective affinities) extends even beyond the vegetable and animal kingdoms. There are not only phylogenetic affinities between animals and plants; there are also chemical affinities by which atoms of bodies of a dissimilar nature unite in certain definite proportions to form compounds. Even if one hesitates to say that hydrogen knows oxygen, both seem to take for granted the reality of the other one. A pointer does not say "game," but it certainly does not direct the hunter's attention to nothing.

In the case of man, everything changes because he is a talking animal. Little children are very curious about animals. As soon as they begin to speak they show themselves remarkable generalizers, and without corresponding classes of objects in nature all generalization would be impossible. Language does not create species; it takes notice of their existence and puts labels on them facilitating their recognition. Long before they can read their names in a picture book, children can tell a dog from a cat in reality.

The same remarks apply even more clearly to adults. It is disconcerting how we can apply the name "dog" to animals as different as a great Dane and a Pekingese, but we do. We even do it quite early in life. I remember a little girl saying that there are three sorts of dogs: tall ones, low ones, and long ones. Adults go further: they distinguish a dog from a wolf, then a wolf from a prairie wolf, as though those animals were not much more alike than dachshunds, greyhounds, and whippets. Except in idealism, the reality of what we call species, or this or that species, precedes our abstract knowledge of them. A particular species can well bear different names in different languages: *canis, chien, dog, Hund, sobak,* and so on, and it is beyond doubt that there exist in reality corresponding classes of beings. When a naturalist takes pride in having discovered a new species, he would feel disappointed to be told that he has simply made it up. During the billion years before man attempted to classify species and to philosophize about them, they were already there, if not in name, at least in reality. The sudden disappearance of all naturalists, even of all men, would leave species undisturbed. Rather, most of them would soon thrive better than before, for man is the most ferociously destructive of all animals. That is what I mean by saying that before being a name, a concept, and a problem, either for science or philosophy, species is an empirically given fact. What does it mean to be "empirically given"? It is simply what William James has called in his *Principles of Psychology* an object of "pure experience,"[8] and which I find described, with more precision than James, by both Aristotle and Thomas Aquinas.

A few lines of Aristotle in the first chapter of his *Physics* show him as embarrassed – as we still are – to give a correct account of our acquiring the different notions of various particular species, and of species in general. Of course, our knowledge of these notions must begin with the senses, but what do we perceive by them: the species or the individuals? This or that particular dog or a member of the dog family? More generally speaking, what do we perceive first: wholes or their parts?

Aristotle answers: neither, and I think he is right. "What is plain and obvious to us at first," he says, "is rather confused mas-

8. [For the notion of pure experience, Gilson refers to William James' *Principles of Psychology* (Chicago, 1952), 852–856. Although the word "experience" is omnipresent in James, the term "pure experience" does not appear in his *Principles*; but he defines it in "A World of Pure Experience," *Journal of Philosophy, Psychology, and Scientific Methods* 1 (1904): 533–543, 561–570. Unlike Thomism, pure experience in James should be understood in the context of his radical empiricism, which "must neither admit into its constructions any element that is not directly experienced, nor exclude from them any element that is directly experienced" (ibid., 534).]

ses, the elements and principles of which become known to us later by analysis." Remarkably enough, the Philosopher then adds that *because it begins with the senses,* knowledge must proceed from generalities to particulars, "for it is a whole that is best known to sense perception, and a generality is a kind of whole, comprehending many things within it, like parts."[9] I said "remarkably enough" because of the trite saying *sensus est particularium, intellectus est universalium.* We only perceive individuals, we only know universals, or, more correctly perhaps, individuals are objects of sense perception, universals are objects of intellectual cognition. And that, of course, is true: a dog is seen or heard; the concept "dog" is neither seen nor heard, it is understood. But we should not infer from the fact that man, who both perceives and knows, or, more exactly, whose cognitions are always at the same time sensible and intellectual, first perceives individuals as such and proceeds from them to his abstract conception of universals. To say: here is a dog, is tantamount to saying: the particular thing which I now see belongs in the species "dog," which I do not see but know. Rather than saying: there is nothing in the intellect that has not first been in the senses, we should say: there is nothing in the senses that is not *at the same time* given in the intellect. Strictly speaking, it is not correct to say "I see a dog," for it would imply that I perceive by the sense of sight the abstract concept "dog," which is knowable by the intellect alone. What I do *see* is a coloured patch which reason knows to be the visible shape of the animal we call "dog."

Now, if we really do not *see* the species, how is it that we all use such expressions as: I see a man, or I see a horse, and so on? The answer of Aristotle and the Schoolmen is well known, if not always well understood. What I perceive by sense is in itself something particular, but my perception of it is something confused. By

9. Aristotle *Phys.* 1.1 184b22–27 [trans. R.P. Hardie and R.K. Gaye, in *The Basic Works of Aristotle*, ed. Richard McKeon (New York, 1941), 219].

observing it more closely and analyzing it, reason forms a clearer notion of it. Seen from a distance, what I see is some thing. If it gets nearer, I see an animal; still nearer, a man. Finally, I see John or Peter. In the end I think I am perceiving by sense, not the sensible qualities of the object, but its very nature. Of course, that is largely an illusion; but there is some truth in it, and in his commentary on Aristotle's *De anima* Thomas Aquinas says why that illusion is justified up to a point. Both the same man, the same soul, perceive by the senses and conceive by the intellect. One should not say that our senses perceive this and our intellect conceives that, but rather that *men* know by both sense and intellect. The two modes of knowledge communicate in the unity of the knowing subject. In Thomas' own words, "Taken at its summit, man's power of sensing somehow participates in that of understanding, because in man sense is conjoined to intellect."[10] In short, because I know that what I am perceiving is a dog, I say I see a dog. In so doing, I merely say that I see what I know I am seeing.

This should be kept in mind every time we wonder about the origin of our knowledge of species. It is nothing scientific; nor was any revelation required to apprise us of the fact that there are species. They are those facts of "pure experience," of which James says, naively but truly, that they "*do stamp copies of themselves within*."[11]

Scripture often says extraordinary things, but always in the language of sense experience, which is but another name for what is often called common sense. Although dramatized and clothed in grandiose poetic language, the biblical revelation of the creation of living species never loses touch with plain reality. It is written in Genesis 2:19–20 that *after* forming every beast of the field and every fowl of the air, God brought them to Adam to see "what he would call them." Thus Adam became the predecessor of Linnaeus

10. [*In 2 De anima* 13.397, ed. Angelo-Maria Pirotta, OP, 4th ed. (Turin, 1959), 101.]

11. William James, *Principles of Psychology*, 861. [James' emphasis.]

and the patriarch of all the classifiers or taxonomists that were to come after him. The names Adam gave to animals are now forgotten. Still, one thing remains the same, namely Adam did not create species, he found them already created. They existed before him, just as they are still given to modern naturalists engaged in the endless task of their classification.

The term "species" is the one used by the Vulgate. God created vegetation yielding seed after its own species: *secundum speciem suam* (Gn 1:11).[12] Then fishes and birds were also created *in species suas* (1:21). In classical Latin, which is a philosophically poor language, *species* means "look," "view," "aspect," "appearance." The original Hebrew word had neither scientific nor philosophical connotations, and I doubt very much that *species* had any such connotations in the mind of St Jerome himself. The King James translation has been well-inspired in resorting to the unpretentious word "kind" to render the Hebrew original. "According to their species" there simply means: according to their kinds. Everybody knows what it is for animals to be "of the same kind." "Kind" is a group of kindred beings. To be created after its own kind is to be created conformable to the type of its own natural group, so as to be at once recognizable as belonging to it.

The species or kind of Genesis is the species given to, in, and by sense experience. The sacred text takes it for granted that everybody knows the meaning of the word. What is at stake in Scripture is not that there are species, but rather that their true origin is the creative power of God. That is neither untrue nor true for science; it is true for faith alone. And the only thing Scripture enjoins us to believe about species, or kinds, or sorts, is the divine nature of their origin. A revealed definition of their nature would

12. [The Vulgate uses both terms, *genus* and *species*. In Gn 1:11 God tells the earth to bring forth fruit trees producing fruit, each according to its own kind (*iuxta genus suum*). *Biblia Sacra iuxta Vulgatam versionem,* ed. Robert Weber (Stuttgart, 1969; 3rd ed. 1983).]

be a great help to philosophers, just as a revealed classification of species would be to zoologists, but no such information is forthcoming from Genesis. We are able to distinguish a horse from a dog or an oak tree from a blade of grass, and all this is purely natural knowledge because we see it. What we do not see, what no person has ever seen, what could not possibly be seen by man, who was a latecomer in the progressive process of creation, is the creative act of God which first caused all beings to be. As for the nature of species, Scripture contents itself with the empirical notion of the realities we designate by that name, and which we unhesitatingly recognize when we see them.

Uncritical as it is, the empirical notion of species nevertheless has several precise characteristics:

a) In the first place, that notion is attended by the immediate and absolute certitude of the reality of its objects. In this respect the real existence of the species "horse," "dog," and so on, enjoys all the privileges of the evidence proper to the objects of sense perception.

The perception of that evidence is accompanied by a feeling of givenness: we do not perceive species as products of our own mental activity, nor, for that matter, of our own practical activity. The French naturalist Lucien Cuénot, in his book *L'espèce*, asks himself why the naturalists make such an extensive use of that notion. As a scientist, he cannot content himself with invoking plain experience to justify the notion. So he finds something else:

> The notion of species results from practical necessity. Man must needs designate by one particular name the beings which he recognizes as different and which he sets apart from other beings. The hunter, the fisherman, the farmer, the gardener have always given group names to the individuals that looked identical to them and which were the objects of their industry or served their own interests.[13]

13. Lucien Cuénot, *L'espèce* (Paris, 1936), 9.

Cuénot's remark is correct if we apply it to the *naming* of the species; but his very description of the process implies that, as was the case with Adam in Genesis, the species must already be there in order to be nameable. Practical usefulness has nothing to do with the existence of species. It is because *there are* different species that we feel the need to give them different names. Hunters, fishermen, lumbermen, gardeners could not give species different names, even for practical purposes, if species were not already there and already perceived as different. Not practical necessity but factual evidence causes fishermen on the St Lawrence to distinguish a pike from a perch, a bass from a muskellunge. Those fishes differ in shape and often in size, hence the diversity of their names. When somebody tells me that there is no muskellunge in big Lake Simcoe, but that there is some in small Lake Couchiching,[14] I may feel surprised, but I know exactly what kind of animal he is talking about. My practical necessity to distinguish muskellunge from other Canadian fishes presupposes the actual existence of such an animal and its distinction from the others in those places. The names matter little. What is called an "achigan" in Québec can be given another name in Ontario ["bass"], but the species remains the same, and its distinct existence is the true reason and cause of the possibility of its receiving a distinct denomination. Without such denominations, botanical and zoological classifications would not be possible. So it is indeed a practical necessity for the naturalist to name species, but his first reason for making such an extensive use of the notion is his empirical certitude of their existence. That is the first remarkable thing about species, that they do exist, and we know it, and neither science nor philosophy can add anything to our certitude about it.

b) This leads us to a second general character of empirically given species. I called its notion "uncritical," and the epithet might

14. [The two lakes are situated to the north of Toronto, Ontario, Canada.]

create a misunderstanding. By that I do not mean that the empirical notion of species is of little cognitive value. On the contrary, it is the only real species there is, because it is the only one that attains species as an objective reality. Far from being unimportant, species for experience is the solid reality that underlies species under all its other forms. We shall have to examine what science and philosophy can say about it, but all that science and philosophy can say about species ultimately rests upon that first kind of species and is related to it. Dobzhansky had in mind those empirically given species when he said: "They do represent tangible biological phenomena. They are facts of life."[15] Facts of life are the same for scientists, metaphysicians, logicians, little children, and in this case, as I said, even for brute animals. It would be less harmful for us to lose all our scientific and philosophical information about the notion of species, including contemporary biogenetics, than to forget the reality of its object.

c) The preceding should settle the question *an sit*, but the question *quid sit* remains untouched, and how can we know *that* a thing is, if we have no idea of *what* it is?

Pure experience asks no question about the kind of reality attributable to species, but since it perceives them as real it must, more or less explicitly, attribute to them at least a certain reality. Leibniz used to say that it is the same thing to say "*one* being," stressing *one*, and to say "one *being*," stressing *being*.[16] And indeed, if a being is not one but two, then we do not have one being but two beings, each of which is one. However, since species are groups, their unity can only be that of a group. As I said, sense

15. Theodosius Dobzhansky, *Heredity and the Nature of Man* (New York, 1964), 100.
16. [In a letter to Antoine Arnauld, 30 April 1687, Leibniz says: "To be brief I hold as axiomatic the identical proposition which varies only in emphasis: that what is not truly *one* entity is not truly one *entity* either." (Leibniz' emphasis.) See *The Leibniz-Arnauld Correspondence*, ed. and trans. H.T. Mason (Manchester, 1967), 121.]

perception does not question itself, but here we find it at the end of its tether; and where empirical evidence ends, science and philosophy begin. What kind of unity, and consequently of reality, can one attribute to a group? For sense experience such a question is out of bounds, and since we are now considering species such as it is for plain experience, we must postpone our examination of the difficulty. But at the same time and for the very same reason, we must imitate the perfect indifference of plain experience to the problem. We must persist, as it does, in attributing to species an actual reality and unity.

d) Experience feels so certain of the reality of species that it attributes to them all the characteristic properties of being *qua* being; that is to say, the properties of all that which is inasmuch as it is.

I have mentioned unity; I must now add self-identity. To be one is to be one with oneself, and this characteristic brings species under the jurisdiction of the principles of identity and contradiction. Like all beings, a species can only be itself. It cannot be itself and at the same time be another species. More simply, no individual can belong to two different species. Naturally, sense experience does not indulge in such abstract reasoning, but it knows that an animal can only be *either* a dog, *or* a mouse, *or* a duck. We cannot even imagine a situation which is otherwise. When he undertakes to describe his imaginary experience of being metamorphosed into an ass, Lucius Apuleius deliberately cheats, for all through his narrative he remains himself, the Roman writer Lucius Apuleius, under the external appearance of an ass.[17] What else could the Latin novelist have done? The *tragelaphos* of Aristotle, the *hircocervus* (that is, the "goatstag") of the Schoolmen was the traditional example in the schools of the impossibility of an animal joining two different species within the unity of its own single being.

17. Apuleius [*The Golden Ass, being the Metamorphoses of Lucius Apuleius*, trans. W. Adlington, revised by S. Gaselee (London, 1915, repr. 1919)].

e) The generality of this belief in the reality of empirical species is confirmed by its tendency to turn itself into a typology. Species are so real for us that each of them is conceived as having a moral character of its own. The severe indictment of the nasty species *cat* in Buffon's *Natural History* bears witness to the depth of these spontaneous convictions. "The cat," Buffon says, "is a faithless domestic animal which one only keeps by necessity."[18] In Walt Disney, who knows his public, bad dogs are as scarce as good cats, and of course the wolf is always "the big bad wolf."

If you feel tempted to consider such remarks trifling, please turn your attention for a moment to the kind of typology widely practiced by nations with respect to other nations, or by races with respect to other races. The temptation to consider groups as actual beings is so strong that we all have heard (not, as I hope, said) sentences of the type: "The English are ... ," "The French are ... ," "The Jews are ... ," as though all the members of one class were susceptible of one and the same definition.

In his *Religio medici*, Sir Thomas Browne denounced that bad habit as an "offence to charity, which no author has written of, and few take notice of; and that is the reproach, not of whole professions, mysteries and conditions, but of whole nations, wherein by opprobrious epithets we miscall each other, and by an uncharitable logic, from a disposition in a few, conclude a habit in all."[19] Incidentally, Sir Thomas Browne did not like Jews.

Sense experience is no infallible judge of moral qualities and defects in the species whose reality it perceives. It can even be wrong in describing them, but it never hesitates about their reality. It now remains for us to investigate, first with the help of science, then with the help of philosophy, the nature of that reality.

18. Comte de Buffon [*Histoire naturelle*, in *Oeuvres complètes de Buffon*, ed. Bernard-Germain-Etienne de la Ville-sur-Illon, comte de Lacépède (Paris, 1825–1828), 12: 412].

19. Sir Thomas Browne, *Religio medici* 2.4, ed. James Winny (Cambridge, 1963), 77.

2. *Species for Science*

In the last lecture I defined species as a fact given in sense experience.[1] Like all the evidences of sense experience, that of the existence of species leaves us in a state of intellectual dissatisfaction. Granted that species do exist, what is the nature of their existence? As St Augustine once said of time, as long as you do not ask me what it is, I know, but as soon as you ask me, I do not know.[2] Still, unavoidably, we will ask. Sooner or later, natural curiosity will prompt us to raise the question: What are species and how are we to conceive them beyond the bare fact that the unlearned are as able, if not more able, than some scientists to recognize them at first sight?

Let us first ask the scientists themselves, what do they call species?

Since we must make a choice, let us begin with the carefully worded definition of Georges Cuvier [1769–1832]: A species is "the collection of all organized beings descended from one another, or from common ancestors, as well as of those that resemble them as much as they resemble one another."[3] Once decoded, Cuvier's definition means that a species consists of plants or animals that are similar in appearance because they descend from common ancestors. And that, a child would say, is the reason daddy bears and baby bears are all called bears.

1. [In the upper left-hand margin Gilson dates the present lecture 15 January 1972.]
2. *The Confessions of St Augustine* [*Confessions* 11.14.17, trans. E.B. Pusey (London; New York, 1907, repr. 1909), 262].
3. Georges Cuvier, *Le règne animal distribué d'après son organisation, pour servir de base à l'histoire naturelle des animaux et d'introduction à l'anatomie comparée*, 2nd ed. (Paris, 1836–1849), 1: 16. [Gilson quotes the definition from Félix Le Dantec in *Lamarck: Pages choisies*, ed. Lucien Brunelle (Paris, 1957), 33.]

Couched in simpler terms than Cuvier's definition, that of Lamarck [1744–1829] is pretty much the same: "*Species* are called all collections of similar individuals produced by other individuals similar to them." To which Lamarck immediately adds, "That definition is exact."[4] And of course it is; individuals belong in the same species because they resemble one another, and they resemble one another because they belong to the same species. Lamarck makes parents resemble their children. Lamarck's definition says something that, ever since the time of Genesis, we have learned from sense experience. Lamarck's science adds nothing to it.

Let us now ask Charles Darwin. In his *Origin of Species* he says:

> Nor shall I here discuss the various definitions which have been given of the term "species." No one definition has satisfied all the naturalists; yet every naturalist knows vaguely what he means when he speaks of a species. Generally the term includes the unknown element of a distant act of creation. The term "variety" is almost equally difficult to define; but here community of descent is almost universally implied, though it can rarely be proved.[5]

Obviously, the whole attention of Darwin is here focused on his own pet problem of the origin of species and varieties. He is particularly anxious to dispel his own former belief in the creation of species by God, as though it made any difference for a species to have been created by God or not. If species have been created by God, that still does not tell us what a species is. And who among us would define a horse: a member of that species that has been created by God such as we now see it? In fact, the remark of

4. Jean-Baptiste Lamarck, *Philosophie zoölogique* (Paris, 1809) 1: 79. [Gilson's translation.]
5. Charles Darwin, *The Origin of Species* [6th ed., 33; Great Books ed., 24. Gilson treats of Darwin's notion of species briefly in *From Aristotle to Darwin and Back Again*, 139–156].

Darwin shows that his concern is the problem of the *origin* of species and varieties, not that of their *nature*. He does not attempt to define species; he does not even pretend to have a personal opinion of the many definitions of it proposed by his predecessors. Darwin merely says that, of those definitions, "no one has satisfied all the naturalists, yet they all know vaguely what it is."[6]

So Darwin cannot quote one satisfactory definition of the fact that is the very subject matter of his book, but he does not worry because he thinks that everybody knows what he is talking about. To the question, how to recognize a species or a variety, Darwin answers as a child would to the question: how does he know a horse from another animal? Well, don't you see it? Of course the species Darwin has in mind are often less well known and harder to recognize, but his answer to the question is the same: ask the specialists! Exactly: "in determining whether a form should be ranked as a species or a variety, the opinion of naturalists having sound judgment and wide experience seems the only guide to follow."[7] Since our guides do not know the scientific definition of what they are looking for, there is little hope for us to find it.[8]

One clear sign of that indifference to the notion is that science does not know how many species there are. Sense experience does not know it either, but that does not worry the layman, because it is a scientific question. Sense experience can count the species it knows, and it does not know more. Here again Genesis is a fair representative of plain experience. God creates herbs, plants yielding seed, and trees bearing fruit after their kind. Later on come birds, fishes, cattle, etc. The question of their number does not arise; we are only told that, whatever their number, *all* kinds of

6. [Darwin, *The Origin of Species*, 6th ed., 33; Great Books ed., 24.]
7. *The Origin of Species* [6th ed., 37; Great Books ed.], 26.
8. [In the margin Gilson adds, "In his recent book *Darwin Retried: An Appeal to Reason* (Boston, 1971), Norman Macbeth pointedly asks in the title of chapter 3, p. 18: 'The Species Problem, or the Origin of What?'."]

plants and animals have been created by God. Every time a naturalist discovers a new species, he teaches us something we did not already know about the world of creation. The more we know about nature, the more we know about God, but plain empirical knowledge of nature, plus, if one cares for it, its poetic knowledge, is good enough for the theologian.

On the contrary, it seems that scientists ought not to content themselves with such a summary treatment of the question. They have taught us to expect from science utmost precision in answers. Yet, if asked how many species there are, they don't know. The question is not like that of the number of stars, which we don't know because most of them escape observation. All the [present] species are actually given and visible to us. We should be able to say, at least, how many of them are now known. But we do not know even that, and the reason for our ignorance of their number is our ignorance of their nature. This, of course, is the main point of interest for us in the strange disagreement among scientists on the question.

Countless examples could be quoted, but *a Jove principium.*[9] In his *Origin*[10] Darwin mentions the fact that, in a case when Mr Babington gives 251 species, Mr Bentham gives only 112 – a difference of 139 doubtful forms. Darwin likewise quotes the remark made by A[lphonse] de Candolle in his admirable memoir on oak trees, that "out of the 300 species which will be enumerated in his *Prodromus* as belonging to the oak family, at least two-thirds are provisional species, that is, are not known strictly to fulfil the definition given above of a true species."[11] But, of course, that definition of a true species by de Candolle has failed to convince other naturalists, and Darwin aptly remarks that "to discuss whether

9. [As Virgil will begin with Jove, Gilson will begin with Darwin. See Virgil, *Eclogue* 3.60; ed. H. Rushton Fairclough (New York; London, 1925) 1: 20.]
10. *Origin of Species*, [6th ed., 37; Great Books ed.,] 26.
11. Ibid., [6th ed., 40; Great Books ed.,] 28.

(certain) forms ought to be called species or varieties, before any definitions of those terms has been generally accepted, is vainly to beat the air."[12]

The situation has not improved since Darwin. One neo-Darwinian, Gandoger, divides the roses of Europe and the East into 5549 species, each of them with a Latin inscription; but according to Lucien Cuénot a recent analysis reduces their number to 211 for the whole genus,[13] a difference, therefore, of 5338 units.

In France, Locard distinguished 251 species of "anodontes" (a kind of fresh-water mollusk), which divide into 19 groups. In Germany, Westerlund recognizes only 87 species of the same mollusks, while according to Schnetter (1922) all the European "anodontes" can be reduced to one single species.[14]

It is the opinion of H.S. Pratt, in his *Manual of the Common Invertebrate Animals*,[15] that there are altogether 822,765 species of animals and 233,000 species of plants, altogether 1,055,765 species of living beings. Now, according to the naturalist Theodosius Dobzhansky, "That these totals fall short of the actually given number of species is clear enough ... . A million and a half species of animals and plants combined is a conservative estimate."[16] In observing such wide discrepancies, let us remember that all those numbers bear upon an undefined object. It is not the ways of counting that differ, it is the nature of the things counted. Even if two of those total numbers happened to coincide, they would be numbers of one does not know what. Naturally this encourages the philosophers to step in and to propose their own total number

12. Ibid., [6th ed., 39; Great Books ed.,] 27.
13. Lucien Cuénot, *L'espèce* (Paris, 1936), 12.
14. Ibid.
15. (Philadelphia, 1935).
16. Theodosius Dobzhansky, *Genetics and the Origin of Species* (New York, 1937), 3. In the third and final edition (1951; repr. 1964), we read instead (p. 8): "A million and a half species of animals and plants combined is, therefore, a minimum estimate of the number now living on earth."

of species. Professor Mortimer J. Adler submits, or rather decides, as the conclusion of an impressive analysis of the notion of the soul, that "there cannot be more than ten species of animate substance."[17] There cannot *be* more than ten; and in fact there *are* only four: man, brute, plant, and body. In his well-known article of 1941 entitled "Solution of the Problem of Species," after carefully reconsidering the question, Adler concludes: "If there are only four proper species, then whenever Aristotle and St Thomas call anything other than *man, brute, plant,* and *body* a 'species,' they must be using that word in a different (and improper) sense."[18] I cannot help wondering at that audacity; I shall content myself with observing that it is a very far cry from Dobzhansky's one million and a half living species to Adler's four.

The naturalists are not scandalized. For them, these wide discrepancies are a normal state of affairs. Theodosius Dobzhansky says that "between a million and a half and two million species of animals and plants have been described since Linnaeus. The exact number is hard to estimate ... . It is considered probable that there may be four million."[19] There is a wide margin between these figures. One reason for it is that the naturalists indifferently use the word "species" to designate any kind of group: species itself, but also variety, family, race, even genus. But a more important reason for that lack of precision is owing to the absence of clear-cut divisions in nature itself. So it becomes a matter of personal choice. In Lucien Cuénot's own words, ever since the time of Linnaeus, there has always been a fight between *lumpers* and *splitters*. To lumpers small differences are negligible; to splitters they are important and should be stressed, thereby multiplying the number of species. Besides, a difference of which one is the discoverer naturally assumes a particular importance. Of the split-

17. Adler, "Problems for Thomists," *The Thomist* 2 (1940): 126.
18. *The Thomist* 3: 350–51.
19. [This sentence is not in the third edition.]

ters, Cuénot says that "they are sometimes right, but if one pursues the analysis he will inevitably arrive at the individual, or at what one can call the pure lineage, and then species vanish."[20]

That excellent naturalist does not seem to realize that here he is at grips with a purely philosophical problem. There is a good reason why one should finally arrive at the individual: there is nothing else in nature; only individuals actually exist. If species are also real, it must be in another sense; they must have another kind of reality.

For species do have a reality, but one feels tempted to say that their reality is perceptible to anyone who is not a scientist, or at least to anyone not looking at them as a scientist. No botanist is satisfied with the scientific distinction of the fir tree and spruce, "two well defined species which woodcutters recognize at first sight."[21] Now, this is the crucial point. What kind of reality can one attribute to a fact recognizable by anyone who sees it, but undefinable for the one who tries to turn it into an object of scientific knowledge? After all, naturalists are as able as woodcutters to tell at first sight a fir tree from a spruce, only the kind of certitude that is good enough for a woodcutter does not set the mind of the botanist at rest. He wants to find *the* character, or characters, *always* present in *all* the individuals of one species and never to be found outside of it. His work proceeds the better as he operates on a smaller number of particular cases. As their number grows larger the exceptions grow more frequent, the transitional forms multiply, the profile of the species gets blurred. Still, while the perplexity of the naturalists grows, their empirical certitude of the reality of species remains unshaken. The way they know them is well described by Lucien Cuénot: "The herborizing botanists are used to the great Linnaean species which one determines with relative facility, and which is known *at a glance*, owing to that

20. Cuénot, *L'espèce*, 13.
21. Ibid., 262.

undefinable ensemble of traits which they call *aspect*."[22] Yes, but there is nothing particularly scientific in that way of knowing things. It is by their aspect that we unhesitatingly tell carrots from cabbages; in doubtful cases we simply learn the differences that enable us not to make dangerous mistakes – for instance, in picking mushrooms.

That ambiguity in the status of the scientific species is reflected in the contradictory statements of famous naturalists. For instance, according to Buffon, in nature there really are only individuals; genera, orders, and classes only exist in our imagination.[23] If this is true, we must be ready to maintain that the species "elephant" or "whale," for instance, taken *qua* species, has no real existence at all. Then why should we give it a name? To Lamarck, "Nature has really formed neither classes, nor orders, nor families, nor genera, nor constant species, but only individuals that succeed one another and resemble those by which they have been produced."[24]

That is perfectly clear, at least as it sounds: species do not exist in nature. But Lamarck's whole book is about species, and his main title to scientific fame is his identification of the large class of the invertebrates, whose very name was coined by him. As for Buffon, in the very same book in which we saw him proclaim that only individuals exist, he says in another passage:

> An individual, whatever its species, is nothing in the universe; one hundred individuals, even one thousand, are nothing. Species are the only beings in nature; those beings are perpetual, as ancient, as permanent as nature itself. They are beings which, the better to judge them, we do not consider a collection or a succession of similar individuals, but a whole, independent of number, independent of time, an ever-living whole,

22. Ibid., 87. [Gilson has condensed a paragraph of Cuénot.]
23. [Quoted by Gilson, in *From Aristotle to Darwin and Back Again*, 39.]
24. [Gilson cites Lamarck's statement in ibid., 166 n. 6.]

> always the same; a whole that has been counted as one item in the works of creation, and which, therefore, constitutes one single unit in nature.[25]

At what moment should we trust Buffon? Are species nothing at all in nature, or are they the only reality there is?

I am far from saying that these contradictions do not make sense. They prove, on the contrary, that, though there is a specifically scientific approach to the description and classification of species, there is no notion of species proper to science. In other words, species for science is the same as species for sense experience; scientific classifications are the scientific ways of handling an empirically given fact. In every naturalist there is a plain layman who knows that there are species just as we do, without being more competent than we are to say in what sense species are realities. The only way we can know what the species "elephant" is, is to see an elephant. But one elephant is not a species, it is an individual; and here we are, caught between sense evidence and science, and tempted to appeal to philosophy as a court of last resort. But to what part of philosophy?

A contemporary school of philosophers who claim Aristotle and Thomas Aquinas for their leaders would unhesitatingly answer: to the philosophy of nature. I have nothing against the project of a philosophy of nature, for there can be a philosophy of everything. Every time you push up to the generalities and principles of some discipline you reach its philosophy, but the philosophy of a discipline is the crowning part of that discipline. So, if there is a philosophy of nature, since the science of nature is physics, that philosophy should be conceived as the crowning part of physics. In fact, it is what Avicenna, and Roger Bacon after him, used to call *communia naturalium*.[26] Such is not the way Mortimer J.

25. [Quoted by Gilson in *From Aristotle to Darwin and Back Again*, 163 n. 12.]
26. [For Roger Bacon's treatment of *communia naturalium* see *Communium natu-*

Adler understands it. To him, the philosophy of nature is "a major division of philosophical knowledge, not only formally distinct from mathematics and metaphysics, but also from all the natural sciences."[27] In this view, the philosophy of nature would be a distinct branch of philosophical knowledge midway between physics and metaphysics.

I have no objection to using the name in order to carry on conversation with my Aristotelian friends, but I find it hard to believe in the reality of the thing. If the philosophy of nature is one major division of philosophy, it should look strange to an Aristotelian that Aristotle himself never thought of it. Yet it is a fact: there is no *è péri phuséos philosophia* in Aristotle, nor is there any *philosophia naturae* in Thomas Aquinas.[28] There is only a physics, including both biology and what we call psychology. Above physics there is only metaphysics, including theology, its crown. To Aristotle, the problem of species belongs in biology, which is one of the main divisions of physics. The supporters of a philosophy of nature are introducing, not only a new division of philosophical knowledge, which would not do any harm, but a new notion of nature, which is a very serious innovation.

According to modern science, the natural practically coincides with the material, which is itself identified by Descartes with geometrical extension in space and the mechanical laws of motion. In a modern treatise of biology, no allusion is ever made to the presence of an immaterial element in living plants or animals. Can you

---

ralium Fratris Rogeri, partes prima et secunda, in *Opera hactenus inedita Rogeri Baconi*, ed. Robert Steele, fasc. 2 (Oxford, 1909). For Avicenna, see his *Liber primus naturalium*, ed. S. Van Riet (Louvain-la-Neuve; Leiden, 1992).]

27. Adler, "Solution of the Problem of Species," 253.

28. [Like Aristotle, Aquinas used the term "natural philosophy" but as a synonym for physics. *In 1 Physicorum* 1.4. See Aquinas, *In octo libros De physico auditu sive physicorum Aristotelis commentaria* 1.1.4, ed. Angelo-Maria Pirotta, OP (Naples, 1953), 14.]

imagine a modern professor of biology beginning his lecture course with a statement like this: "No part of an animal is purely material or purely immaterial." Well, this is a quotation from Aristotle.[29] If we wish to update Aristotle and Thomas Aquinas, we naturally must add, owing to their insufficient information about nature, the recently acquired scientific information about the living cell and its nucleus, its genes and its chromosomes, but all that leaves intact the traditional notion of zoology as one part of physics. Cells, chromosomes, and genes are considered as purely material elements, so that biology remains the physics of living beings. In order to constitute a philosophy of nature midway between modern physics and classical metaphysics, we would have to deny the presence of an immaterial element in any one of the parts of living natures, including cells, chromosomes, genes and all the rest. In short, we must first deny the Aristotelian notion of nature.

This is what modern scientists do, with the result that strikingly evident biological facts become unintelligible to their very science. In my lectures on final causality in 1970 I observed that all biologists, including Darwin and his present-day successors, are unable to account for final causality and deny its presence in living organized beings. All they can do about it is to call it "teleology" or "teleonomy" instead of using the discredited term "final causality." But changing names does not change things. This year we are also observing that scientists do not know what to do with the concept of species, which is both evident to sense experience and incomprehensible to science.

I beg to suggest that in the two cases the nature and cause of the difficulty are the same: scientists are mistaking a philosophical problem for a scientific one. Will you forgive me for over-stressing my point? It is not simply that on their own admission naturalists,

29. Aristotle *De partibus animalium* 1.3 643a25 [trans. William Ogle, in *The Basic Works of Aristotle*, ed. Richard McKeon (New York, 1941), 653].

and particularly taxonomists, cannot agree on a common definition of species. The point is that, while confessing their inability to define species, they continue to count them, to classify them, and even to investigate their origin, which is properly to look for the origin of one-does-not-know-what. When taken together the two facts are remarkable. As Lucien Cuénot says in the very first sentence of his book *L'espèce*: "The question of species shows the paradoxical feature that everybody uses the name without defining its meaning."[30] To that I simply add that everybody uses the name because at the level of sense experience everybody knows what it is, and no scientist cares to define it because "species" is not properly a scientific notion.

But perhaps what science cannot do can be done by philosophy. In the last of these lectures I shall attempt to define the kind of philosophical knowledge we can have, and indeed do have, of species, conceived as resulting from the presence of an immaterial element in material reality.

30. Cuénot, *L'espèce*, preface.

3. *Species for Philosophy*

We now turn to philosophy for an answer to the question: What is a species? The answer must satisfy three evident data of pure experience:

1. A species has no actual existence of its own; only individuals actually exist;
2. A species is an evident reality given in nature;
3. A species is even much more real than any one of the individuals it comprises.

Buffon, Lamarck, Darwin, and other scientists have confirmed the validity of this empirical data, but neither science nor plain common sense has done anything to reconcile the contradictory elements it contains, namely, only individuals enjoy actual existence; species, not individuals, are the true realities of nature.

As we appeal to philosophy for an answer to this riddle, to what part of philosophy should we direct our attention? The first one we might think of is logic, but logic is the last part of philosophy from which we can hope to receive an answer to the question. Logical relations are relations between concepts of more or less generality, and the more general is related to the less general as genus to species. Given any genus, it suffices to limit its generality by a specific difference in order to obtain a species. "Animal" is a genus, virtually containing many species. One of them is obtained by adding to the genus "animal" the difference "rational"; "rational animal" is the definition of the species "man."

This is simple and clear. The trouble is that the logical species is not the one whose nature we are trying to understand. For science as well as for experience, species are realities given to sense perception, whereas logic is in no way concerned with factual realities. It only deals with concepts in the mind and sets up the rules that preside over the correct way to associate them in judgments, reasonings, and deductive and inductive demonstrations.

Pure logic is no more concerned with actual reality than mathematics; neither one is "ontological." The Schoolmen were well aware of the fact. The logician, Thomas Aquinas says, is concerned with the mode of predication and not with the existence of a thing: *Logicus considerat modum praedicandi et non existentiam rei.*[1]

Unfortunately, ever since the time of Porphyry and Boethius, logic has been taught not as a distinct discipline to be studied in itself but as an introduction to philosophy, with the disastrous result that all the subsequent sciences have been both studied and taught as if they were applications of the methods of logic to the various orders of reality. I said that the results have been disastrous, because the quasi-sterility of science in the medieval schools is mainly due to this error.

Had they been more attentive to the teaching of their master, the ancient Thomists would have copied in golden letters this simple line: "the mode of the thing understood in understanding is other than the mode of the thing in being" (*Alius est modus intellectus in intelligendo quam rei in essendo*).[2] The two orders of knowledge and reality are totally different in nature: our concepts are immaterial, the things conceived are material. So, as Thomas says, our intellect understands material things in an immaterial way; and again, "its mode of understanding material things is immaterial."[3]

How does this affect the problem of species? In a very direct way. The material beings that we know are individualized by their very matter; they are all individuals. In other words, only individuals exist in reality. This is the *modus rei in essendo*: individuality, caused by matter, is the thing's mode of being. On the contrary, since the intellect is immaterial nothing can enter it with-

1. *In 7 Metaph.* 17.1658, ed. M.R. Cathala, OP, and Raimondo Spiazzi, OP (Turin, 1950, 1964), 396.
2. *Summa theologiae* 1.13.12 ad 3 [Dominican college ed. (Ottawa, 1941–1945), 1: 90b].
3. Ibid.

out losing its individuality; so that while the thing known is always an individual, our knowledge or concept of it is always a universal. Materiality and individuality, immateriality and universality always go together. This is the fundamental reason the sciences of life are bound to get in trouble sooner or later: they only deal with individuals and they must conceive them by means of universal concepts. Hence, in the biological sciences [there is] an inevitable gap between their knowledge and their object.

As an example, let us consider the problem of definition. In logic it is simple. In the classical definition of man, "rational animal," the genus "animal" is specified by the difference "rational." Scholastic masters and students without number have been trained to mistake that logical definition of man for his biological definition.[4] But it is not. A rational dolphin would be a rational animal but it would not be a man. The first two chapters of Aristotle's admirable *De partibus animalium* contain an explicit demonstration of the impossibility of classifying animals by the simple method of dichotomy. You cannot have all the animals on

4. I find it difficult to agree with men whose opinions carry much weight with me: "[T]he notion of species is strictly a logical concept ... any use of the word 'species' to signify anything other than the concept *species* itself is a derivative mode of signification. Strictly speaking, the concept *species* is never used ontologically; the word 'species' can be used ontologically" (Mortimer J. Adler, "Solution of the Problem of Species," *The Thomist* 3 [1941]: 298 n. 27; see John N. Deely, "The Philosophical Dimensions of *The Origin of Species*," *The Thomist* 33 [1969]: 77 n. 4). "[T]he 'species ontologically considered' is not exactly the 'specific nature' but rather an application of the logical notion of species to designate the specific nature" (Jacques Maritain, "Concerning a Critical Review," *The Thomist* 3 [1941]: 46 n. 3; see Deely, ibid., 77 n. 6). Yes, but where did Aristotle himself find his own logical notion of species? The naturalists he has so closely studied, and whose methods of classification he so often criticizes, were not indebted to any logic for their notion of species. Logic does not come first in the history of that notion. But it is too true that the Schoolmen have mistaken the logical species of Aristotle for those of zoological classification and, by their teaching of logic, caused countless others to make the same mistake.

one side and rationality on the other. The rational animal we call man is one very particular kind of animal: he is a mammal, a primate, a biped, endowed with speech and therefore with reason, unless you prefer to say: endowed with reason and therefore with speech. This description remains very incomplete, yet, unlike the mere word "animal," it conveys at least a sketchy notion of its object. The way man is in reality is quite other than the way the concept "man" is in our understanding. Generation after generation of Schoolmen have mistaken the order of concepts in the mind for the order of things in reality.

The difficulties that beset the problem of species have the same origin. Since they are concepts in the mind, genera and species are immaterial; hence they are universals and, by the same token, they are not real beings. For a species to be a real being, it would have to be at the same time both an individual and a universal: a universal in order to be a species and an individual in order to be a being. This is a point on which no follower of Aristotle and Thomas Aquinas should entertain the slightest doubt. To the question: are universals actual beings? the answer is, no. This is even the reason why, unlike Plato, Aristotle does not admit of universal causes (that is, of universals that are actual causes). Universal causes do not exist because universals do not exist. It is always an individual that is the originative principle of another individual. "There is no universal man, but Peleus is the originative principle of Achilles, and your father of you, and this particular [letter] *b* of this particular [syllable] *ba*," though *b* in general is the [originative] principle of *ba* in general, but only a real *b* can be the principle of a real *ba*.[5] Again, in *De partibus animalium* Aristotle says, "The individuals contained within a species, such as Socrates and Coriscus, are the real existences."[6]

5. Aristotle *Metaph.* 12.5 1071a19–22 [trans. W.D. Ross, in *The Basic Works of Aristotle*, ed. Richard McKeon (New York, 1941), 876].
6. *De partibus animalium* 1.4 644a22–23 [trans. William Ogle, in *The Basic Works of Aristotle*, 655].

Here, however, very few Thomists have followed Thomas Aquinas to the very end. They seldom do. If Thomas says that the *quid est* of God is wholly unknown to us (*penitus ignotum*), they say that we know it a little.[7] If Thomas says that essences are unknown (*essentiae sunt ignotae*), they say, yes, but not quite.[8] Thomas does not really mean it. If you tell them that Thomas denies all actual existence to universals, the same Thomists attribute to him what they call a "moderate realism." But the expression is not to be found in the writings of Thomas Aquinas himself, and besides it is meaningless, for a thing can do anything in moderation except to be. To be moderately real is not to be real at all; it is to be nothing.

Thomists have only one answer to this conclusion: if Thomas is not a realist, not even a moderate one, then he must be a nominalist, which we know he never was. In fact, the opposition between realism and nominalism does not fit the problematic of Aristotle and Thomas Aquinas. Thomas never said that universals are realities, nor did he ever say that they are only names. He would not say that species are either *things* or *names*. The answer to the question must be looked for in another direction, not that of logic but of metaphysics. We must look at that which, in a memorable volume, Aimé Forest has aptly called "the metaphysical structure of reality."[9]

Let us try to ask the question in the same terms as Thomas Aquinas himself. To him, as to Aristotle,[10] substance is actual real-

7. [For Aquinas' expression *penitus ignotum* see *Summa contra Gentiles* 3.49.8.]
8. [For Aquinas' expression *essentiae sunt ignotae* see *De veritate* 10.1 resp.]
9. Aimé Forest, *La structure métaphysique du concret selon saint Thomas d'Aquin* (Paris, 1931).
10. This is the meaning of the distinction introduced by Aristotle between primary substances (actually existing individuals) and secondary substances (genus and species) (*Cat.* 5 2a12–18). The terminology is misleading, for secondary substances do not subsist, but the ambiguity is meaningful, because, while they do not subsist at all by themselves, genus and species *are*

ity. Only substances are actual beings; all the rest only exist inasmuch as they belong to substance and share in its being. Since we cannot help trying to imagine what we think, let us imagine substance as one, solid, ontological block, cast in one piece along with all its determinations; for instance, each and every one of us. All that which we are: faculties, operations of body and mind, all our knowledge, all our habits of intellect and will, everything in us is

---

the substance, unless one prefers to say that the substance *is* the genus and the species. Genus and species are not accidents of the substance, they are it. The relation *Socrates is man* differs entirely from the relation *Socrates is white;* Socrates could not be white, he could not be man without ceasing to be Socrates. Two statements make the meaning of Aristotle quite clear: (1) Unlike accidents, species are not present in their subjects, they are only predicable of them: "Of things themselves some are predicable of a subject and are never present in a subject. Thus 'man' is predicable of the individual man and is never present in a subject" (ibid., 2 1a20 [trans. E.M. Edghill, in *The Basic Works of Aristotle,* 7]). To say that Socrates is not predicable and to say that he is the real substance is to say the same thing; to say that species is predicable and that it is present only in the mind, is to say the same thing. (2) On the other hand, since they are formal constituents of the individual, only species and genus deserve to be called secondary substances, "for these alone of all the predicates convey a knowledge of primary substance" (ibid., 5 2b30–31 [trans. Edghill, in *The Basic Works of Aristotle,* 10]). Because genus and species are but predicates, they do not subsist; when Aristotle speaks of secondary substances as those *within which* primary substances are included (ibid., 5 2a15), he means to say: in the mind. Perhaps the most perfect formulation of the doctrine runs as follows: "All substance appears to signify that which is individual. In the case of primary substance this is indisputably true, for the thing is a unit. In the case of secondary substances, when we speak, for instance, of 'man' or 'animal,' our form of speech gives the impression that we are here also indicating that which is individual, but the impression is not strictly true, for a secondary substance is not an individual but a class with a certain qualification; for it is not one and single as a primary substance is; the words 'man,' 'animal,' are predicable of more than one subject" (ibid., 5 3b10–18 [trans. Edghill, in *The Basic Works of Aristotle,* 12]). Where the received translations say *substance,* the Greek says *ousia,* that is, beinghood, entity, the very stuff beings are made of. See Aristotle *Metaph.* 7.16 1040b5–1041a6 [trans. Ross, in *The Basic Works of Aristotle,* 809–810].

our very self and only exists as belonging to the one substance that we are and in virtue of its own act of being. Thomas says, "all that which is, is by its own act of being: *est per suum esse*."[11] Such ontological blocks are individual by definition. Consequently a collective entity such as a species cannot be a substance, which means that, taken in itself, it cannot exist. On the other hand, it can exist *in* substance as one of its constituent elements. Endowed with no *esse* of its own, a species shares in the *esse* of the substance, the more so as the substance cannot be a substance without it. As Thomas says with Aristotle, substance cannot be predicated of anything, but other things can be predicated of it.[12] One cannot say that man is a Peter or a John, but one can say that Peter or John is a man. Species is of that sort: humanity is attributable to John; it has no actual existence of its own, but it shares in the actual existence of each and every human being.

The next question is: What is there in every substance that is individual, as substance itself is, and nevertheless, because it itself is not a substance, is predicable of many (which is to be a universal)?

That specific element itself cannot be a universal. To say that universals are in substance and specify it, would be to forget the distinction we made between the way concepts are in our understanding and the way things are in reality. As you will recall, their *modus essendi* is not the same. Universals are not to be found in things but only in minds. In the century-long strife between realists and nominalists, the mistake of the realists was precisely to speak of "the reality of universals," as though in order to exist universals had to be things. The relentless criticism of that kind of realism by the nominalists consisted in showing that a species, like man, horse, and dog, could not be at one and the same time,

11. [*Summa contra Gentiles* 1.22.5., ed. Peter Marc et al. (Turin, 1961–1967), 2: 32, no. 206.]
12. Aristotle *Cat.* 5 2a11–33; Thomas Aquinas, *In 5 Metaph.*, 10.898, ed. M.R. Cathala, OP, and Raimondo Spiazzi, OP (Turin, 1950, 1964), 241.

either in whole or in part, present in a multiplicity of individuals. If it were a thing, a species would be one substance, and then it would not be a species but an individual.

The most perfect formulation of that criticism is to be found in William of Ockham, who is justly considered the best representative of nominalism. Ockham is so precise on this point that it is a pleasure to listen to him:

> If someone says that Socrates and Plato agree in reality more than Socrates and an ass, therefore Socrates and Plato agree in something real in which Socrates and an ass do not really agree. Now, since they neither agree in Socrates nor in Plato, they must agree in something else, in some way distinct from both and common to both. – To this I answer that, strictly speaking, it ought not be granted that Socrates and Plato agree *in some thing*, nor in *some things*, but that they agree *by themselves*, and that Socrates agrees with Plato, not *in something* but *by something*, namely by himself.[13]

One cannot go further in denying the reality of universals in things: "Socrates and Plato really agree by themselves more than Socrates and an ass, not however in something real."[14] If one feels satisfied with that answer to the problem, there is nothing more to say. Ockham sees Socrates, who is a man, then he sees Plato, who is also a man. For the two to agree, each one of them only has to be that which he is. Ockham knows what he is talking about, and he is maintaining a pure philosophical position from beginning to end.

So the difficulty remains. What is there in Socrates that is also in Plato but is not in a donkey, and which is real without being an individual, that permits us to say that both Plato and Socrates are men? I find the metaphysical answer to the question in the[15] inex-

13. [*In*] *Sent.* 1.2.6, ed. Stephen F. Brown and Gedeon Gál (St Bonaventure, N.Y., 1970), 2: 211–212.
14. Ibid., 212.12–14.
15. [Gilson first added 'philosophically,' then deleted it.]

haustible *Summa theologiae* of Thomas Aquinas. After recalling that a universal is neither an actually existing principle nor an extramental being, Thomas adds: "But if we consider *the generic or specific nature itself as existing in individual things* – [let us memorize these words] – then *in a way* it plays the part of a formal principle with respect to the individuals, for the individual is owing to matter, while the notion of species comes from form."[16]

So universals have no existence outside of the mind, nor have forms existence outside of things, but forms have reality in the things whose actual being they cause and share. To Ockham, Thomas Aquinas would reply that Plato and Socrates do indeed agree by themselves, but only because they both agree in the form of humanity. And how can two distinct individual substances agree in one and the same form? Only because form is an immaterial principle. Being an immaterial element in substance, it can be participated by any number of material individuals. A form is real in a real being, but it itself is not a being. Form is universal in the mind that abstracts it from matter, yet it is not a universal present in things. "Species is from form,"[17] Thomas says. Can we, living in an age of plain scientific materialism, still take that notion seriously?

16. *Summa theologiae* 1.85.3 ad 4. That is why I find it difficult to admit that "the notion of species is strictly a logical concept." The concept *species* is never used ontologically as implying separate actual existence; but, apart from its logical use, in which it truly signifies nothing but itself, species points out the secondary substance, that is to say, the formal determination whereby the primary substance actually is that which it is. The reality of species (as actualized in and by the primary substance) is such that Aristotle dares to write: "Of secondary substances, the species is more truly substance than the genus, being more nearly related to primary substance" (*Cat.* 5 2b7 [trans. Edghill, in *The Basic Works of Aristotle*, 9]). It is not easy to understand the plus or minus in the category of substance as a merely logical relation – not, at least, in the doctrine of Aristotle, where forms are the acts of substances and their formal causes.
17. *In 1 Perihermenias* 8.96, ed. Raimondo Spiazzi, OP (Turin, 1955), 39; *Summa contra Gentiles* 2.72.3 [ed. Peter Marc et al., 2: 206–207].

It depends on whether or not we can still accept the presence of an immaterial element in physical reality. I quite agree that very few modern naturalists will consider it a possibility. Descartes' devastating criticism of substantial forms has eliminated them from the world of science. Boileau and Molière have ridiculed them. Today, scientists do not even go to the trouble of rejecting them; their very notion has been forgotten. To Descartes, animals were not material bodies organized by immaterial forms; they were material machines. Even the human body was nothing else. Descartes has carried the day, because it was [through] working on his assumption that modern science, particularly modern biology and physiology, has achieved and is still achieving its marvelous progress. Mechanicism is indeed completely justified where living organisms are, in fact, organized material machines; but mechanicism falls short of an answer on two points. First, when asked to answer the question, what is matter? Next, when asked to assign the organizing power to which these machines owe their existence. Neither can it find an answer when asked for the principle of unity present in all the individuals of the same species. So why not try again the old answer of Aristotle which I have already quoted: a species is constituted by a combination of form with matter, "for no part of an animal is purely material nor purely immaterial."[18]

If we agree to try this answer, of course what is immaterial in a material thing can only be its form. I quite agree that nothing immaterial can be taken into account by today's *science*, but it can be taken into account by *philosophy*, one of whose most important tasks has always been to consider those parts of nature for which science has no explanation.

Such is the case with the problem of species. Form is not a scientific solution to it, but it is a philosophical notion suggested by the scientific data of the problem. The need of such an appeal

18. Aristotle *De partibus animalium* 1.3 643a25 [trans. Ogle, in *The Basic Works of Aristotle*, 653].

to philosophy is beginning to make itself felt in some scientific quarters. After all, it cannot be pleasant for a scientist like Darwin to write a book on *The Origin of Species*, feeling all the time that one does not know what a species is, or like Lucien Cuénot to entitle a book *Species* only to conclude that there is no quite satisfactory definition of it.

Studying the kindred notion of race in his book on *Heredity and the Nature of Man*, Dobzhansky says, "The question of the *reality* of races and species has been a sort of biological football for a long time."[19] A very apt comparison indeed, except that in the case of species it is a football game in which there is no ball. Everybody wonders where it is and even if it exists.

With great acumen, the same biologist adds that, in fact, the species game is "really a part of an even older philosophical football, the dispute between the nominalists and the realists," the nominalists reducing species to group concepts "residing in the minds of those who invent them." Resorting to another comparison, the same biologist adds that, to a nominalist, "classification of organisms is like that of postage stamps, to be arranged as conveniently and effectively as possible."[20] And here again the comparison is excellent, except that, once more, in a postage stamp collection the stamps are real and their classification by countries of origin is quite natural. No stamp collector has ever doubted the reality of his stamps or the objective validity of their classification, which is not the case with living species.

The point I am striving to make is that there is a basic fault with biological classifications, either animal or vegetal, but that this is not a fault in scientific method. Rather, the fault consists in looking for a scientific answer to a philosophical question. Perhaps philosophy can find the answer; perhaps philosophy has already

19. Theodosius Dobzhansky, *Heredity and the Nature of Man* (New York, 1964), 104.
20. Ibid.

found it. But of course philosophy cannot have found a *scientific answer*, from which scientists will naturally conclude that no answer at all has been found. I think it has, and the answer is, form.

First, the immateriality of form accounts for the possibility of groups of beings, both individually distinct owing to their respective matters, and specifically alike owing to their forms. Where every other explanation succumbs to the criticism of Ockham, this one stands.

Second, hylomorphism fully agrees with the feeling, often expressed by naturalists, that, while in a way only individuals exist, in another sense nothing matters in nature but the species. We have seen Buffon declare that, though the species has no actual existence, it is the true biological reality. Now this is also true if living beings are endowed with forms, for it is a principle of Thomas Aquinas that "the ultimate intention of nature is towards the species and not the individual or the genus, because the form is the end of generation, while matter is for the sake of form."[21]

Incidentally, that speculative view is supported by countless facts: first of all, by the colossal waste of seeds and the high mortality rate of the young in all species of living beings, including man. The waste is real from the point of view of the number of individuals, but nature is willing to sacrifice millions of them in order to ensure the survival of the species.[22] [Second], in men, the old instinct of "save the women and children first" still witnesses to the inborn conviction of the primacy of species over individuals. Third, the immaterial nature of substantial forms answers

21. *Summa theologiae* 1.85.3 ad 4 [Dominican college ed., 1: 529b–530a].

22. The amount of that waste is beyond imagination. According to Lucien Cuénot (*L'évolution biologique; les faits, les incertitudes* [Paris, 1951], 367), salmon lay 10,000–20,000 eggs every year; herring 30,000–70,000; shad 100,000–150,000; cod 5–7 million; and turbot 5–15 million. On the contrary, the hardy and voracious little stickleback is said to lay 200 to 300 eggs, and that only twice in a lifetime.

the traditional objection that they are but names and that, anyway, we know nothing about them.

In Molière's *Le malade imaginaire*,[23] the play ends with the grotesque examination in dog-latin of a candidate for the degree of Doctor of Medicine. The president of the jury asks the candidate to quote the cause and reason why opium puts one to sleep:

> Domandabo causam et rationem quare
> Opium facit dormire.

To which the candidate answers: because there is in opium a dormitive virtue whose nature it is to put the senses to sleep:

> *To that* respondeo:
> Quia est in eo
> Virtus dormitiva
> Cuius est natura
> Sensus assoupire.

Of course that is ridiculous. And is it not just as ridiculous to say that an animal is a lion because it has a substantial form whose effect is to make it a member of the species Lion, a *Felis Leo*?

Yes, that is ridiculous, too, if we are looking for a *scientific* answer to the question: What is a lion? But there is no scientific answer to the question precisely because the nature of species is not a scientific problem. A notion is scientific if it permits us to foresee and to predict. If it is true, it says how certain things come to pass. Perhaps some of you will remember the ill-tempered remarks of Charles Darwin about Herbert Spencer that I quoted in last year's lectures.[24] They were the typical reaction of the scientist

23. Molière, *Le malade imaginaire*, in *Oeuvres complètes*, ed. Maurice Rat (Paris, 1956), 2: 906.
24. [That is, the lectures he gave on final causality in 1970. Darwin's remarks about Spencer are in *The Autobiography of Charles Darwin, 1809–1882*, ed. Nora Barlow (London, 1958), 108–9. See Gilson, *From Aristotle to Darwin and Back Again* (London, 1984), 68.]

to the philosopher who contents himself with submitting an intelligible description of reality such as it is: That man has never helped me to predict anything!

It is true that a substantial form does not enable us to foresee and to predict anything. We posit it as an essence, but we don't know it except as the cause of its effect. We do not know the form whereby a man belongs to the species "man"; nor do we know the form whereby any other animal or plant belongs in its respective species. On the contrary, we know the species from our knowledge of individuals. As Thomas Aquinas says, "The essences of things are unknown to us, but their powers are revealed to us by their acts."[25] Now, since the specific differences are also powers known to us only by their acts, we know the acts, not the specific differences. So we can name species and affirm their reality on the basis of their forms, but the forms themselves are only known to us by their effects.

To conclude, there is only one single kind of real species, namely that of sense experience, which is the apprehension by reason of an empirically given fact perceived by sense. Language itself suggests this conclusion, for when we say "a dog", "a" points out one concrete individual, while "dog" points out an abstract species. Every individual is given in a species; every species is given in an individual. It is the privilege of experience to perform such epistemological wonders without being aware of it.

The enduring validity of Aristotelianism is due to its decision to stand by common-sense knowledge and to abide by its decisions. Both science and philosophy are critical reflections on common-sense knowledge, but on different levels.

Science engages in an inventory of the data of experience. In the present case science pursues an objective description of vege-

25. ["[R]erum essentiae sunt nobis ignotae, virtutes autem earum innotescunt nobis per actus" (*De veritate* 10.1 resp., in *Quaestiones disputatae*, ed. Raimondo Spiazzi, OP (Turin, 1949), 1: 191).]

tal and animal species. It teaches us that some fishes are mammals, that there are four-footed reptiles, and other such information that escapes the uncritical perception of the senses. That is why science is so often critical of sense, because, while accepting the reality of its object as a real fact (that is, there *are* species), it does not trust the way the senses apprehend it. Philosophy also trusts the testimony of the senses as to the reality of its object, but it does not proceed, like science, to a detailed inventory of its objects. Philosophy simply wants to know how such an object is possible. In other words, supposing the senses are right, what is it necessary to admit for it to be right? In the case of species it is necessary to admit, in matter, the presence of an immaterial element that is the form of that matter. Only such an immaterial element can account for the formal unity of groups in a world where only individuals really exist. That is the abiding meaning and truth of the well-known scholastic doctrine called "hylomorphism," which does not belong to science, as the moderns understand it, but to philosophy.

In what part of philosophy should it be located? In physics, as I said; but to be in the physical world, as we are, does not mean to be in the purely material world. What the Schoolmen used to call "material forms" were so called because such forms could not subsist apart from matter, not because they were matter. Since subsisting immaterial forms are the proper objects of only philosophical[26] and theological knowledge in traditional philosophy, the consideration of living natural species is certainly physical, but it belongs in a physical knowledge from which the metaphysical is not excluded. Here again, the division of sciences is no satisfactory pattern for a division of nature. Thomas says: "Inasmuch as bodies are composed of matter and

26. [The philosophical knowledge Gilson has in mind here is metaphysics.]

form, they approach the divine likeness because they have form, which Aristotle calls something divine."[27]

"Something divine": such is the form, any form *qua* form. In such a philosophy our choice is not between physics and metaphysics, but rather between materialized modern science and Christianized Greek nature, in which matter never exists without a form, that is to say, without an immaterial element, one of the intelligible forms radiated by the Divinity.

E.G., Cravant (Yonne), revised 27 August & 13 September 1971

27. Aristotle *Phys.* 1.9 192a16; Thomas Aquinas *Summa contra Gentiles* 3.69.27 [ed. Peter Marc et al., 3: 98].

III

In Quest of Matter

Etienne Gilson

Introduction

In Cravant, on 30 July 1972, Gilson finished his preparation of his next three lectures for Toronto, "In Quest of Matter." The papers surveyed three ways of examining matter: the Greek method, the Christian philosophers' method, and the scientific method followed since Descartes.

For the Greek approach to matter, Gilson looked at Aristotle's *Metaphysics*. He called the Greek approach "matter for sense experience," and argued that the Greeks were not interested in submitting a thing to calculation and measurement for possible practical uses. They wanted to understand it as it struck them, and then, if possible, to define it. Matter to the Greeks was the obscurity out of which a thing was produced by a form. Matter was a mode of being; it was not nothing but was an incomplete, a potential being.

The Christian philosophers presented matter through a variety of depictions. Augustine called it, beautifully if inadequately, the mutability of mutable things. Albert the Great called it an "order toward a form," a kind of halfway post between nonbeing and being. Thomas called it "being *in potentia*," a could-be-but-not-yet being. Some scholastics, for example Scotus, saw matter as a *res*; they located this *res* in nature however, where it could be made the object of a science.

Although Descartes and his scientific successors have presented many varying opinions on matter, most have thought of matter as a thing discoverable in its properties and uses, and as useful and profitable to science. For them matter is no longer an obscurity, a mutability, or a potency; rather, matter is an extension, a quantity, a figure or situation, a movement. For the sciences the distinguishing mark of matter is its usefulness. The ancients and the medievals asked themselves the philosophical question, "What is matter?" The moderns merely pry into its properties.

> The result of Descartes' victory is that modern scientism leaves us without any metaphysical notion of matter … . Science has discovered an unbelievable quantity of truth concerning the structure of material bodies; it has pushed beyond the notion of atom and discovered in elementary molecules an infinitely small universe of unbelievable complexity. … Because [scientists] are busy investigating the nature of matter, they believe they are examining the question of what it is.[1]

In Gilson's view, despite all such progress, "the nature of reality is other than our knowledge of it." ... The philosophy of Aristotle and the scholastics was always concerned with reality as perceived by the senses and translated into such terms of conceptual knowledge as being, cause, becoming, quantity, quality, relation, and so on. The peculiar satisfaction the mind derives from matter arises from reality as given to the senses, and is as obscure as reality itself. As presented by the philosopher, matter is only one of all those philosophical translations of sensory experience that science spurns as incapable of quantification within clear and distinct categories. Philosophy prefers to keep reality as it truly is – to suppress reality's obscurity is to remove one of its essential parts.

As Gilson sent these three lectures to the Institute, he felt his teaching had reached some point of completion:

> I probably would have modified them if I had felt able to deliver them, but the substance would have remained the same. Anyway, these were the last lectures I intended to give. They deal with the last of the metaphysical concepts I had wanted to clarify in my own mind. In a sense, the thought satisfies me. To realize that the moment has come when I no

1 [Here Father Shook quotes several passages from Gilson's lectures on matter. For the passages in question, see below, p. 118.]

longer have anything personal to say leads me to hope that there was something in these lectures. I felt an urge to deliver them before the end.[2]

Laurence K. Shook[3]

2. Gilson to Shook, cited in Shook, *Etienne Gilson* (Toronto, 1984), 389.
3. Taken from his *Etienne Gilson*, 388–389.

1. *Matter for Sense Experience*

Two years ago, in 1970, I devoted four lectures to the notion of final causality. In my lectures of 1972 I attempted to clear up the notion of species.[1] Now I wish to submit the notion of matter to a similar philosophical examination. My reason for making this choice is that the notion of matter is the last of a few [philosophical notions] about which I have found it difficult to satisfy myself. By the word "to satisfy," I do not mean: to form a clear and distinct understanding of the notion but, rather, at least to see why such an understanding of it is not possible.

In my philosophical examination of those notions, I have kept that of matter for the end, because it has always seemed to me to be the most baffling. It is, I think, entirely foreign to the modern scientific mind. Contemporary scientists never think of it in terms of its past philosophical meaning. I shall endeavour to show why, but I do not feel sure to succeed in the undertaking.

It is always possible for one scientist to understand a definition of matter set forth by another scientist. For instance, Leibniz has exactly interpreted the notion of matter propounded by Descartes, as we shall have occasion to observe. But the reason for that possibility was precisely that Descartes' notion of matter was not philosophical but scientific in nature. On the contrary, it is very difficult to make a scientist understand the philosophical implications of the philosophical notion of matter. The Aristotelian notion of matter looked ridiculous to Descartes because it was of no scientific use. It is my intention to go back to the basic philosophical notion of matter and to make a special effort to understand the purpose it seems to have served in the mind of the ancient philosophers.

1. [That is, the three lectures above, pp. 33–74.]

The Greeks never tried to develop philosophy from science but from common-sense experience. They were not chiefly interested in submitting matter to calculation and measurement in view of some possible practical uses. Rather, they were anxious to understand its nature, [in order] correctly to see what it was and, if possible, to define it.

Matter – in French *matière* – are translations of the Latin word *materia,* which meant in classical Latin, and in what I believe to have been its first meaning, *wood.*[2] When Cicero wrote: "omnis materia culta et silvestris,"[3] he meant: all sorts of wood, cultivated as well as wild. In Livy, *materiam caedere* means to fell trees.[4] And indeed, wood was the very type of that of which a thing is made. But by a most natural extension of meaning, wood has become the name for what all material things are made of – then, by a further extension, the constituent substance of all things, made or not made.

The Greek word for matter is *ulè,* the root of all our words beginning with *hylo,* for instance *hylophágous,* that is, wood-eating (like some insects), or else, by a similar extension of meaning *hylozoism,* the doctrine that all matter has life. In short, the evolution of the Greek *ulè* and of the meaning of the Latin *materia* is the same. Starting from wood, the material *par excellence,* it ends with what a thing is made of in general.

It is noteworthy that, exactly as in the case of final causality or purposiveness, the origin of the notion of matter lies in the domain of making. Now to make is to cause something to be, or

2. [For *materia,* see *Harper's Latin Dictionary: A New Latin Dictionary,* ed. E.A. Andrews, Charlton T. Lewis, and Charles Short (New York; Oxford, 1879), 1118.]
3. Cicero *De natura deorum* 2.60.151 [in *De natura deorum: Academica,* with trans. by H. Rackham] (London; New York, 1933), 268.
4. Livy [*Ab urbe condita* 5.55.3, in *Livy,* with trans. by B.O. Foster et al., (London; Cambridge, Mass., 1919–1959), 3: 186].

rather to become, and to become implies the notion of growing to be from something else, of changing or developing into something either by spontaneous self-development or in consequence of the action of an external cause. Something cannot be made out of nothing; every recipe to make something begins with the formula: in order to make something, first take this or that, as the case may be, as in the threadbare French joke: "Pour faire un civet, prenez un lièvre" ("To make a jugged hare, first catch a hare").[5] What strikes the attention of the common-sense observer in every case of becoming is that *something* is becoming or is caused to be. In other words, every becoming is the becoming of something.

That is a primary evidence which cannot be denied; but what is evident is not always clear. It may be evident that it is so, without the nature of the fact in question being clear. On the contrary, I feel tempted to say, from bitter philosophical experience, that the nature of evident facts is often wrapped in obscurity. Practically all metaphysical notions are of that nature: fundamentally they point out facts known by immediate sense experience and are incapable of translation into terms of complete rational intelligibility. Such are all facts involving the notion of becoming, as is here the case.

In order to come to be something other than that which it is, a thing must be both itself and something else. How can that which is, at one and the same time be and come to be? The fact that what the thing comes to be is something does not help. On the contrary, there's the rub! If there is becoming, which nobody can doubt, it seems to disagree with the principle of identity or of non-contradiction, which is one of the first principles of reason.

In all cases when we are at grips with such primary notions, if we want to philosophize about them, the last resort I know is Aristotle. I know the Philosopher is not in fashion, but I also am of

5. [The translation is Gilson's. We would say today, "To make rabbit stew, first catch a rabbit."]

the opinion that philosophy has little to do with fashion, because, as I understand it, philosophy is interested in true knowledge, and unless there is reason to think that its object itself has changed in the course of time, why should one suppose that true knowledge of it has also become different? More simply, if there is no reason to think that becoming itself has changed since the time of Aristotle, why should one *a priori* disregard as outdated what the Philosopher has said on the question?

In point of fact, I see at least one reason to pay great attention to everything Aristotle said on philosophical questions, especially on questions related to principles. The philosophy of Aristotle rests upon indestructible foundations, because its principles are primary evidences, immediately certain to sense experience, or to reason, or to both. On the other hand, nothing is weaker than evidence because, *qua* evident, it evades demonstration. Blaise Pascal has forcibly stressed the point that one of the causes of scepticism in philosophy is the ambition to demonstrate principles.[6] Since the thing cannot be done, those who attempt it simply conclude that, since they are not demonstrable, principles are not true. The Philosopher himself never made that mistake. To him, principles were primary, self-evident facts, whose proper function was to justify, not to be justified. Contemporary philosophy is no longer interested in accepting facts, because that is realism. So philosophers rather try to justify principles by way of dialectical deduction, with the result that they have no use for Aristotle. But when someone happens to see principles in the same positive factual way, he recognizes the Philosopher as a philosophical brother. Aristotle then comes back to life. I fancy Oliver Wendell Holmes had something like that in mind when he wrote in his *The Autocrat of the Breakfast Table* that Aristotle was a "regular dandy"[7] – not

6. Pascal, *Pensées* 1.6.101, ed. Michel Le Guern (Paris, 1977), 1: 104–105.
7. "There was Aristoteles, a very distinguished writer, of whom you have heard, – a philosopher, in short, whom it took centuries to learn, centuries to

indeed that Aristotle is never out of fashion because truth has nothing to do with it.

In *Physics* 1.7, Aristotle approaches the problem in a roundabout way, "with reference to becoming in its widest sense." For, "we say that one thing comes to be from another thing, and one sort of thing from another sort of thing."[8] Leaving aside unnecessary distinctions – I mean, unnecessary for our own purpose – we shall say that "from surveying the various cases of becoming ... there must always be an underlying something, namely that which becomes, and that this, though always one numerically, in form at least is not one."[9] We speak of coming to be, or of coming to be so and so. The first case applies to substances, which "are only said to come to be in an unqualified sense."[10] In all other cases, a substance is said to become so and so. The substance is the subject that becomes; but even substances come to be from some underlying substratum, namely from that which becomes a substance. And this is the point where Aristotle, for want of a satisfactory definition of that substratum in general, lists a series of cases in which becoming manifestly proceeds from an underlying subject:

> Generally things which come to be, come to be in different ways (1) by change of shape, as a statue; (2) by addition, as things which grow; (3) by taking away, as Hermes from the stone; (4) by putting together, as a house; (5) by alteration, as things which 'turn' in respect of their material substance. It is plain that these are all cases of coming to be from a substratum.[11]

---

unlearn, and is now going to take a generation or more to learn again. Regular dandy, he was." Holmes, *The Autocrat of the Breakfast Table* (London; New York, 1906), 249.

8. *Phys.* 1.7 189b34 [trans. R.P. Hardie and R.K. Gaye, in *The Basic Works of Aristotle*, ed. Richard McKeon (New York, 1941), 230].
9. Ibid.
10. Ibid., 190a33 [trans. Hardie and Gaye, in *The Basic Works of Aristotle*, 231].
11. Ibid., 190b5–8 [trans. Hardie and Gaye, in *The Basic Works of Aristotle*, 231].

One could not keep closer to sense experience and to its common-sense interpretation: in every case of becoming, there is something that becomes, some subject or substratum (*upokeimenon*) that first was under one form and then is under another form. That same subject plays the part of matter in the process of coming to be. Aristotle does not find it easy to define "subjectibility" as such, and yet that is the only common character proper to all subjects *qua* subjects. Here again, in order to achieve some degree of concreteness, the Philosopher resorts to a series of comparisons:

> The underlying nature is an object of scientific knowledge, by an analogy. For as the bronze is to the statue, the wood to the bed, and the matter and the formless before receiving form to any thing which has form, so is the underlying nature to substance, that is, to the *this*, or existent.[12]

So far, all we can say of matter is that it is "the formless" before it has received its form. But why do we say *its* form? Because no matter is susceptible of receiving all sorts of forms. Bronze, which is the matter of the statue, has the form of bronze; wood, which is the matter of the bed, has the form of wood. It is only with respect to the bed that wood is formless; it lacks the form of the bed, just as bronze is formless with respect to the statue until it has been moulded into one. In themselves, bronze and wood are complete substances, but they are matters with respect to that which can be made out of them. With respect to the forms they can still receive, they are in a state of privation. By "privation," Aristotle does not simply mean something which a certain matter does not have but, more precisely, something that matter is susceptible of having, and which it ought to have in order to become another sort of thing. Such is the meaning of Aristotle's cryptic formula, that the principles of all becoming are

12. Ibid., 191a7–11 [trans. Hardie and Gaye, in *The Basic Works of Aristotle*, 232].

three in number: matter, privation, and form.[13] A process of becoming has been completed when a certain matter has received a certain form with respect to which it was in a state of privation. The matter (the subject matter) is in potency to the form before receiving it; it is in act with respect to the form after receiving it. The passing of a certain matter from potency to act is a case of coming to be or be-coming.

Let us try to simplify this, for though its verbal expression is unavoidably complex, the notion itself is simple. To say that matter is being in potency seems contradictory because, in itself, being is act, and act is being. To be in act, actually to be and to be are one and the same thing. How then can something be, that is, be in act, and at the same time be in potency? To the question, the Philosopher's answer is that being in potency must be something possible since it is real. The only condition for its possibility is that a being be not both in potency and in act at the same time and in the same respect. Now our experience of becoming is of beings that are, that is, are in act (otherwise they would be nothing at all), and yet, because they are finite beings, are still able to add to their own being, or to complete their own being by acting upon themselves.

Finite being is not only a fact; it is the only kind of being given in experience. To know why it exists raises the problem of creation; to say how it can exist is to raise the thorny problem of essence and existence, since it is because it has a definite essence that a being is finite. In short, we are here confronted with a cluster of the most formidable metaphysical problems. The problem of matter and its relation to form, or of potency to its relation to act, is but one particular case of the same fundamental problem, which is that of the possibility of finite being. To philosophize like Aristotle requires a spirit of submission to reality. I do not think I am exaggerating in saying that humility is at the very origin of

13. [Ibid., 1.7 191a20, trans. Hardie and Gaye, in *The Basic Works of Aristotle*, 233.]

realism. I feel tempted to add that pride is at the very origin of idealism. At any rate, idealism defines reality such as it ought to be in order to be fully intelligible. The trouble is that, in order to be fully intelligible, the laws of nature ought to be identical to or, at least, of the same structure as those of the mind. The Hegelian temptation would then be justified: philosophy, physics, even history would be as many branches of logic. Reality would be a sort of Mind; it would be the object of a possible mental construction, the dreamland of a university professor of philosophy.

One has to make a choice: either logic is an organon, an instrument, and of course a science, but a science of the functioning of the mind, as is the case with Aristotle, or else reality itself is a dialectic, in which case the science of the functioning of the mind becomes, by the same token, a science of reality. The common root of modern idealism, under its various forms, is the Cartesian ambition to build up a philosophy exclusively containing "clear and distinct ideas." And, of course, not only mathematicians but all men delight in clear and distinct ideas. The question is: does reality itself consist of clearly and distinctly definable elements? What is typical of philosophical realism is to accept reality such as it is, even confused and obscure, because it is not certain that things can be given in the mind under the same form as they subsist in reality. The chief reason for the difference is that, in minds, things subsist under the form of immaterial concepts, whereas objects always contain a material element in physical reality. Objects cannot carry their matter into immaterial minds; they must, so to speak, leave matter at the door, so that we cannot hope to confer upon them the intelligible qualities of clarity and distinctness proper to concepts. Hence the seemingly verbal character of many Aristotelian notions and definitions. They do not aim to achieve clearness but to adhere to the often involved nature of reality.

In describing it,[14] Aristotle uses at least two sets of terms: potency and act, and matter and form. A being that is changing in any way becomes in act something it was in potency. We are not here speaking of the active power to act, which Aristotle and the Scholastics often mistakenly call active potency, but of the passive potency, or aptitude to be acted upon and to be actualized by some cause. There is a reason for that ambiguity, namely the fact that there must be in passive potency a positive aptitude to be acted upon. In Aristotle's own words:

> Obviously, then, in a sense the potency of acting (that is, power) and of being acted on is one (for a thing may be "capable" either because it can itself be acted on or because something else can be acted on by it), but in a sense the potencies are different. For the one is in the thing acted on; it is because it contains a certain originative source, and because even the matter is an originative source, that the thing acted on is acted on, and one thing by one, another by another; for that which is oily can be burnt, and that which yields in a particular way can be crushed, and similarly in all other cases. But the other potency is in the agent, for example, heat and the art of building are present, one in that which can produce heat and the other in the man who can build.[15]

Paper *can* be folded, stone cannot. The positive character of passive potency is also well known and appreciated by artisans or artists who use various matters or materials. Quite a few years ago, when I asked an American businessman what was his line, he answered: copper. Well, I said encouragingly – I was mentally comparing it with metaphysics – is that good? Well, the man answered, "It used to be, but now the future is in plastics." My fellow traveller was doubly right. First, at that date the industrial

14. [Gilson first added "as he sees it," then made a correction that is illegible.]
15. *Metaph.* 9.1 1046a19–27 [trans. W.D. Ross, in *The Basic Works of Aristotle*, 821].

possibilities of plastics were beyond imagination; second, the answer was quite Aristotelian, for the meaning of *plastic* is: that which can be moulded, shaped, or formed. (Another name for passive potency could be plasticity.)

Such is generally the case with the principles of Aristotle. At first sight they express in needlessly abstract terms [elements] borrowed from plain sense experience. On closer inspection it appears that plain sense experience itself is something very complex that cannot be described in simple terms. Two years ago [in the lectures of 1970] this appeared to us to be the case with the notion of final causality, which describes the present as already predetermined by its own future. Then in our lectures of 1972 we dealt with the notion of species in plants and animals, which have no real existence except as individuals, and yet share in common the characters of a certain group. This year[16] our quest of matter leads us to the conclusion that reality consists of opposite elements, each of which can only be described as the contrary of the other: form and matter, act and potency, and so on. Let us read from the *Metaphysics*:

> *Actuality*, then, is the existence of a thing not in the way which we express by *'potentially'*; we say that potentially, for instance, a statue of Hermes is in the block of wood, and the half-line is in the whole, because it might be separated out, and we call even the man who is not studying a man of science, if he is capable of studying; the thing that stands in contrast to each of these exists actually.[17]

To the objection: very well! but all that does not give us a definition of act and potency or of form and matter, it only points them out. Aristotle's reply should be kept in mind by all those who, like

16. [If Gilson had delivered these lectures it would have been in 1973.]
17. *Metaph.* 9.6 1048a31–35 [trans. Ross, in *The Basic Works of Aristotle,* 826. Gilson's emphases].

myself for one, intend to keep faith with philosophical realism: look at similar cases, compare them, and just say what you see:

> Our meaning can be seen in the particular cases by induction, and we must not seek a definition of everything but be content to grasp the analogy, that it is as that which is building is to that which is capable of building, and the waking to the sleeping, and that which is seeing to that which has its eyes shut but has sight, and that which has been shaped out of the matter to the matter, and that which has been wrought up to the unwrought. Let actuality be defined by one member of this antithesis and the potential by the other.[18]

This prepares us for Thomas Aquinas' terse definition of matter: "materia ... est ens in potentia."[19] First, matter is *being*. To the extent that it is being, it must, at least up to a point, be act. It is something in act, that is, under some respect, in potency with respect either to itself (I can learn) or to something else (I can change place in space). In both cases, let us remember that every time we say form or matter, act or potency, the notion we have in mind is always actual-being, potential-being, formal-being, or material-being. *And* would be better than *or*, since, as we have seen, one of those modes of being, or, as Thomas sometimes calls them, one of those substantial modes, can only be understood in contradiction with the other.

Nothing could show better how foreign to the modern and scientific notion of matter that traditional and philosophical notion is. We ask Aristotle: What is matter? He answers: It is a certain mode of being. This is so true that, in the priceless philosophical lexicon of Aristotle, *Metaphysics*, Book 5, one vainly looks for a chapter on matter. What information there is on the subject is found

18. Ibid., 1048a36–1048b5 [trans. Ross, in *The Basic Works of Aristotle*, 826].
19. [*Summa theologiae* 1.66.1 ad 3, Dominican college ed., 1: 402b; *In 1 Physicorum* 14.8, ed. Angelo-Maria Pirotta, OP (Naples, 1953), 65 no. 257.]

between chapter 6 on *One* and chapter 8 on *Substance*, in chapter 7 on the various ways things are said *To be*. After listing several senses of the word, Aristotle adds, at the end of the chapter:

> Again "being" and "that which is" mean that some of the things we have mentioned "are" potentially, others in complete reality. For we say both of that which sees potentially and of that which sees actually, that it is "seeing," ... . And similarly in the case of substances; we say that the Hermes is in the stone, and the half of the line is in the line, and we say of that which is not yet ripe that it is corn.[20]

Another way of reducing the notions of potency and of matter to that of being is to recall that, in all orders, act is prior to potency:[21] "For from the potentially existing the actually existing is always produced by an actually existing thing, e.g., man from man, musician by musician; there is always a first mover, and the mover already exists actually."[22] In short, "everything that is produced is something produced *from* something and *by* something, and that the same in species as it."[23] The something from which the substance is produced, whatever it may be, is matter; that by which it is produced is the form; and because it is a substantial form, it produces the substance in the very same species of which it is the form.

20. [*Metaph.*] 5.7 [1017a35–b8, trans. Ross, in *The Basic Works of Aristotle*, 761. In his later, revised translation Ross changed the first sentence here to read as follows: "Again, 'being' and 'that which is,' in these cases I have mentioned, sometimes mean being potentially and sometimes being actually." *The Complete Works of Aristotle: The Revised Oxford Translation*, ed. Jonathan Barnes (Princeton, N.J., 1984), 2: 1606].
21. Ibid., 9.8 1050a2–3.
22. Ibid., 1049b23–26 [trans. Ross, in *The Basic Works of Aristotle*, 829].
23. Ibid., 1049b27–29 [trans. Ross, in *The Basic Works of Aristotle*, 829. Gilson's emphases].

It now remains for us to see what that notion of matter has become in the doctrines of the great medieval Aristotelians, Albert the Great and Thomas Aquinas.

2. *Matter for Christian Philosophers*

The Latin word *materia* occurs only a few times in the Latin Vulgate, in three cases with the meaning of "wood" as a building or carving material.[1] The word does not appear in Genesis; it makes no allusion to matter as something created in the beginning by God.

The word entered theological speculation when Christian thinkers with a philosophical formation introduced it in their commentaries on Scripture. Every time they did, those theologians naturally gave the word the meaning it had in the doctrine of the philosopher they had studied. For instance, after giving up the materialism of his youth, chiefly under the influence of Plotinus, Augustine conceived all things as made up of both matter and form. When Scripture says that in the beginning God made heaven and earth, the "heaven" spoken of is a spiritual matter already formed, namely angels. The reason Augustine posits the existence of spiritual matter is that, following the tradition of the Platonists, a purely spiritual and immaterial being would be a god. In Augustine's mind, materiality and mutability are inseparable. God has immutably fixed the angels in their beatitude, but like all creatures they are naturally mutable. It is to their spiritual matter that they owe their mutability.

It goes without saying that *a fortiori* corporeal creatures also are made up of matter and form, and here again matter is the principle of mutability. Or is it? In order to be a principle, matter must be something. Now we cannot think of a thing without conceiving it as defined by a certain form, and matter can have no

1. [The Douay-Rheims translates *materia* in 4 Kings 6:2, 5 as "timber"; in Wisdom 11:18 it translates the phrase *quae creavit orbem terrarum ex materia invisa* as "which made the world of matter without form"; *terrae materia* in Wisdom 15:13 is rendered "earthy matter."]

form. The great merit of Augustine is to have forcibly stressed the two main characters of the notion of matter: to be both inevitable and inconceivable. In *Confessions* 12.3–9, he reminds his Christian reader that, when God first created it, "the earth was invisible and unorganized" (Gn 1:2). This means that before God formed and distinguished into things of different kinds that unformed matter, "there was not any definite thing, neither colour, nor shape, nor body, nor spirit. Yet it is not that there was absolutely nothing: it was a certain formlessness without any species."[2]

Obviously, this is still far from satisfactory. What is a formlessness? We naturally imagine something with a distorted form, or with countless successive forms, but that is not to be formless. To deny its existence would be easier than to "think of something in between form and nothing, something neither formed nor nothing, an unformed thing which is almost nothing."[3]

The gist of the difficulty obviously lies in mutability itself. Directing his attention to bodies themselves, and "working more deeply into their mutability, by means of which they cease to be what they were, and begin to be what they were not," Augustine came to think that the transition from form to form "was accomplished by means of some unformed thing and not by means of complete nothingness."[4] But it does not escape Augustine that he still leaves us with the unsatisfactory notion of a *formless thing*. He then tries another answer: it is the very *mutabilitas* of things which is their matter: "The mutability, then, of mutable things is itself capable of receiving all the forms into which mutable things are changed."[5]

2. *Confessions* 12.3.3. [Gilson uses the translation of Vernon J. Bourke, *The Essential Augustine* (New York, 1964), 106.]
3. Ibid., 12.6.6, [trans. Bourke, *The Essential Augustine*,] 107.
4. Ibid.
5. Ibid.

That is a very clever way out, but nothing more, for what is that mutability? Is it a thing, or is it the property of a thing? Augustine himself is baffled. What is this? he asks: mind, or body, or any species of either one? Were I permitted to say it is a "nothing-thing," or an "is-is not," I should say it. Still, he lamely adds, "it would have to have some kind of being in order to receive the visible and organized forms."[6] Finally Augustine gives up. What he is anxious to make clear is that, whatever matter is or is not, it has been created by God "in the beginning." The personal experience of Augustine, and its final failure, is far from meaningless. It sets into relief the mysterious nature of a notion which thinkers in the Middle Ages were to find as baffling as Augustine himself had already found it to be.

This, however, must be said in favour of Augustine, that he had perfectly localized the difficulty. For it to be intelligible, matter must be conceived as a being that is not, which of course is unintelligible, but it is only unintelligible because it identifies the two notions of being and thing. Let us remember the formula of *Confessions* 12.6.6: "If one could say 'nothing-thing' and 'is-is not', I should say, this were it." To him thing and being were equivalent, as it still is for us when we define a thing as a reality. And here is an excellent opportunity to understand the reason the Schoolmen have preferred Aristotle to Plato. Aristotle has not turned matter into a translucid reality, but he has pointed out the direction in which philosophical thought can hope to find it. One thing at least is sure, it is not nothing, so it must be some sort of being: not a thing, but a certain way or manner of being. And this takes us back to the terse formula of Thomas Aquinas: matter is being in potency. Both Albert the Great and Thomas Aquinas himself have done their utmost to make us realize the meaning of Aristotle's answer to the problem.

6. Ibid.

Matter, Albert says, can be neither defined nor described, because it is absolutely simple, and all definition or description of an object consists in enumerating its parts. Since it is being itself, matter shares in the property of being not to be a genus specified by a difference. If we merely want to describe an object, we must at least enumerate its chief characteristics. After quoting some examples of descriptions, Albert simply concludes: "Ergo descriptio non potest esse materiae, cum sit simplex" ("Therefore there cannot be a description of matter, since it is simple").[7]

We naturally feel tempted to infer from what precedes that matter is simple, in the sense that it is a simple thing, or a thing that is simple. But that description of matter would fit God, who is an absolutely simple being and yet is supremely immaterial. No, we are looking for matter in those finite things we call material. It is not their substantial forms, nor is it their accidental forms; but, since we cannot conceive it in itself, we must try to conceive it in relation to form or in conjunction with form. Only thus will it become somehow definable, because the part of the definition provided by the form will be susceptible of definition. To repeat, matter is not a thing in the proper sense of the word: *non est absolute res*; it is a thing relatively. "In truth, it is an order toward form, which order is between being and non-being."[8]

To be an "order toward form," to be something between being and non-being, does not amount to being much of anything. On the other hand, there is some encouragement in the fact that Albert's puzzlement is very much the same as Augustine's. The notion of matter looks equally obscure to both of them; but at least it displays the same kind of obscurity to them: one cannot say of it

7. Albert the Great *Summa de creaturis* 1.2.2 [*B. Alberti Magni Ratisbonensis episcopi, ordinis Praedicatorum, Opera omnia*, ed. Auguste Borgnet (Paris, 1895), 34: 323a].

8. "[E]t est secundum veritatem ordo ad formam, qui ordo medium est inter ens et non ens." Ibid., 1.2.4 solutio, [*Opera omnia*, ed. Borgnet, 34:] 330a.

that it *is* fully, nor that it *is not* at all. Albert, moreover, realizes that our failure to define it is owing to the fact that we try to define it in itself, while it probably needs to be defined in its relation to something else, namely to the thing to which it is ordained.

A first remark is that, just as an operating power is not disposed to perform any kind of operation, so also a passive possibility is not disposed to become any conceivable kind of being. It is therefore a sensible question to ask of any given thing: what is it capable of becoming, or what can possibly be made out of it? In one of his fables La Fontaine depicts a perplexed sculptor asking himself before a block of marble: "What will it be, a god, a man, or a washbasin?"[9] He does not ask himself if he will turn the marble into a piece of real string or a loaf of real bread. There are things which any given object is susceptible of becoming and others which, of their very nature, they are incapable of becoming. That very disponibility in view of a certain ulterior state, that natural fitness for becoming this and that other kind of thing, constitutes, for Albert the Great, the very essence of materiality. In order to be material, then, a thing must both be lacking something and be able to have it. Using an Aristotelian metaphor, Albert says of the thing in question that it "desires" the perfection it is susceptible of having and has not; it is "deprived" of that perfection. God, who is infinite and perfect being, is everything it is possible to be. As he is lacking in nothing, there is nothing he can possibly become. Consequently God is totally immaterial. On the contrary, every finite being is lacking something precisely inasmuch as it is finite. Its finiteness, which consists in that which it could be *in its own order*, and is not, is not a mere absence of being, it is a privation. Now to be deprived of is to desire. To desire is to experience the

9. [*The Fables of La Fontaine* 9.6, "The Sculptor and his Jupiter," trans. Marianne Moore (New York, 1964), 214.]

privation of something one feels able to have. That ability to have something of which one feels deprived, is matter itself.[10]

All that is very abstract, but it is far from meaningless. Matter, we said, is something between being and non-being. According to what has just been said, matter is first something negative: it is a lack, a want, a deficiency, a shortcoming, and all that stands on the side of non-being. On the other hand, matter is also a desire to fill up that gap along with the capacity to do so, and this stands on the side of being. Now, to be able to become a certain thing is not to be that thing, and yet it is not nothing. It is, in the very nature of a being, what enables us to judge, as the saying goes, if one will be able to make something of someone: a champion in this or that particular sport, a singer, a writer, a teacher, and so on. We then say of a certain subject that he has "the makings" of one or the other of these walks of life. There is good "stuff" in him. But inorganic materials also have their own possibilities, their aptitudes, as we said, to become this rather than that. To repeat, that aptness to become is not a being, yet it is not nothing. In industry, the discovery of a "better material" to produce something can be the starting point of an industrial revolution.

An immediate and important consequence of this way of understanding matter is that, in this view, matter cannot exist apart. Since it is not an *aliquid*, but a mere receptibility of form, its existence is only that of a subject in potency to the perfection it is susceptible of receiving. This is even true of so-called "prime matter" (*materia prima*). They call prime matter the subject which is left after being stripped of all its forms. Rather, let us say, of all its forms but one, for a total absence of form would be a mere nothingness. So there must always remain at least one form, namely the form of quantity.[11] By the same token, there is no

10. "Ad id quod contra ..." [Albert the Great *Summa de creaturis* 1.2.4 solutio, *Opera omnia*, ed. Borgnet, 34:] 330.

11. "[M]ateria prima nunquam est sine quantitate ... est semper conjuncta quan-

matter common to all beings. Since the matter of a certain being is its aptitude or aptness to receive a certain perfection it is lacking, that matter is proper to it. In short, matter is always a certain matter; matter in general does not exist.

As to its essentials, the doctrine of Albert will remain that of his disciple Thomas Aquinas. Strictly speaking, our imagination somewhat exaggerates the distance that separates the master from the disciple in time. The life of Thomas Aquinas is wholly contained within the limits of that of Albert the Great who, born before Thomas (1193), died six years after him (1280).[12]

Without pretending to give a precise and detailed account of the possible reciprocal influence between the two theologians, I shall content myself with saying that, on the whole, the two doctrines constitute a fairly homogeneous doctrinal bloc. Passing from Albert to Thomas, one rather feels like [he is] following one and the same thought striving to achieve more and more precision in expression.

In both doctrines it is impossible for matter to exist apart. Not all the Schoolmen agreed on that point. Some of them would say that, since it has been created by God, matter must be a certain thing. On the contrary, in a philosophy where being is act, by definition the notion of a pure potency, like formless matter, endowed with actual being, is meaningless. This is so true that, according to Thomas, even God could not create prime matter deprived of all form, because that which is without form is without actuality and therefore without being. It is against the nature

---

titati isti vel illi." Ibid., [*Opera omnia*, ed. Borgnet, 34:] 330b.

12. A similar illusion of perspective occurs in the case of Franz Joseph Haydn, whom we always imagine as an ancient with respect to Mozart, even though the life of the younger (1756–1793) falls entirely within the life of the elder (1732–1809). N.B. The composer of *The Creation* survived the composer of *Don Giovanni* by 16 years. [Gilson puts this remark in the body of the lecture].

of matter, not that it be created but that it be created without form.[13] It is, let us say, "concreated" along with the form. Again, "All that which is found in nature exists in act. Now matter only exists in act through form, which is its act. Consequently no matter without form can be found in nature."[14]

Let us go back to our starting point: matter is essentially being in potency: *materia, secundum id quod est, est ens in potentia.*[15] This entails one further consequence of great importance, namely that matter is always that of a certain particular being. It is the matter proper to that being to which it is in potency. Contrary to the opinion of many Schoolmen, there is no "form of corporeity" (*forma corporeitatis*) common to all bodies. The form of each particular body, by which it is constituted as distinct from the others, is its own form of corporeity, and that body needs no other one.[16] Matters of the same species are matters actuated by forms of the same species, for of itself no matter is like or unlike any other one. The more completely a matter is actualized by its form, the less it is able to change by receiving other forms. There is one being that is pure form, pure act, and therefore it is totally immutable. That being is God. But if there is a Pure Act, there is no Pure Potency, no Pure Matter. By the same token, as every being is act, even its matter is that of a determined being.

The contrary illusion is due to the notion of prime matter, which is that of matter in general. It is the notion common to all particular matters, actual or possible, but that notion is not that of a real matter common to all corporeal beings. It is only an abstrac-

13. *Summa theologiae* 1.44.2 ad 3.

14. *De potentia* 4.1 resp. [ed. Paul Pession, in *Quaestiones disputatae*, ed. Raimondo Spiazzi (Turin, 1949), 2: 104–105].

15. *Summa theologiae* 1.66.1 ad 3. [Dominican college ed. (Ottawa, 1941–1945), 1: 402b; *In 1 Physicorum* 14.8, ed. Angelo-Maria Pirotta, OP (Naples, 1953), 65 no. 257.]

16. *Summa theologiae* 1.66.2 ad 3.

tion, without any reality beyond that of a concept in the mind, that is, of a being of reason, not of a real being.

In reality, therefore, real prime matter cannot be pure formlessness. It consists of the elements, themselves composed of elementary qualities combined in certain proportions. In other words, really existing prime matter is the four elements (fire, air, water, and earth), themselves composed of a certain proportion of the elementary qualities: heat, cold, dryness, and moisture. These elements played, in the doctrine of Aristotle and the Schoolmen, the same part as the atoms in all atomistic physics. And, unlike the abstract notion of prime matter, they were very real and actually existing beings. The coming to be of every being consists in a certain arrangement of those elements under the action of an efficient cause and following the pattern of some substantial form. Its passing away occurs when the form allows a physical being to dissolve into its elements. During that process of dissolution, the material body successively loses all the properties it owed its substantial form, but it is never annihilated. There always remains something, that is, some material element determined by its elementary forms. In the words of the sixteenth-century French poet Ronsard: "*La matière demeure et la forme se perd*" ("The matter remains and the form gets lost").[17] When Gustave Cohen read that verse, that worthy historian of French literature rejoiced. Here, he said, is a perfect case of Renaissance Epicurean materialism.[18] Alas! that was only Ronsard's recalling the purest Aristotelian scholasticism. St Thomas says:

> All the creatures of God in some respects continue forever, at least as to their matter, since what is created will never be annihilated, even though it be corruptible ... . For corruptible

17. Pierre de Ronsard, *Oeuvres complètes*, ed. Gustave Cohen (Paris, 1950), 2: 118.
18. Gustave Cohen, *Ronsard, sa vie et son oeuvre*, 5th ed. (Paris, 1956), 284.

> creatures endure forever as regards their matter, though they change as regards their substantial form.[19]

The matter that perpetually remains is not, of course, formless prime matter, which we have just shown does not exist; rather, it is one of the four elements or any combination of them.[20]

We have still to consider the part played by matter in physical change. Change is the passing from one definite state to another. That transition (*transitus*), or the passing from a certain condition to another, implies the existence of a certain thing that is the same in the two extremes of the change, transition, or motion. The subject of change must be the same at the beginning and end, though it is not to be found in the same condition at those two moments. The same water becomes either steam or ice, the same runner is at the starting point and at the goal of the race, the same man is first an embryo, then a child, an adult and, finally, an old man. All that which was in the ore, then in the molten metal, is still to be found in the bronze statue, and it will outlast the statue if it happens to be broken into pieces or even molten and liquefied by fire. At each moment of some such transformation, or series of transformations, that which is becoming must be *able to become* or *to be turned into* something else. That ability to become, or to be caused to be, necessarily lies in a certain being: it is the very aptness to change which we called potency and which we identified with matter when we said that matter is being in potency (*ens in potentia*). The subject common to the two extremes of a change is not a being in act; it is only a being in potency.[21]

It could still be said that in every process of becoming matter is *that which* becomes. A matter is always a certain thing. Thomas has carefully read book 12 of Augustine's *Confessions*, where mat-

19. *Summa theologiae* 1.65.1 ad 1 [in *Basic Writings of Saint Thomas Aquinas*, ed. Anton C. Pegis (New York, 1945), 1: 610].
20. Ibid. [Gilson's reference; but the term "four elements" is not found there.]
21. *De potentia* 3.2 resp.

ter is defined as mutability itself.[22] And indeed, mutability is the characteristic property of all matter, but matter is always the mutability of something, of a certain thing. And that precisely is the reason a wholly formless prime matter is impossible. To be formless is to be without act, hence without being. As was said, there is to be found in reality a prime act without any potency, but not a prime potency, not perfected by some act. And for this reason there is some form even in prime matter.[23] More correctly, let us say: in one prime matter, or, in that which is the prime matter of a certain thing, for every being has its own prime matter, which is not a mutability but a mutable thing.

Hot, cold, dry, moist, hard, and soft are elementary qualities of matter perceptible to the senses. Inasmuch as it is the subject of these elementary qualities, sensible matter is called corporeal matter. But corporeal matter has still other properties, not sensible but intelligible, which are the mathematical properties of quantity. There is therefore an intelligible matter, whose form is quantity.

In a very remarkable passage of the *Summa theologiae*, Thomas begins by distinguishing between matter in general and individual matter; let us say: between flesh or bone in general and this bone or that piece of flesh. Then he continues: "Mathematical species, however, can be abstracted by the intellect from sensible matter, though not from common intelligible matter, but only from individual matter, that is, from physical beings either perceived or imagined." Here is the essential part of this passage:

> For sensible matter is corporeal matter as subject to sensible qualities, such as being cold or hot, hard or soft, and the like, while intelligible matter is called substance inasmuch as it is

22. See above, pp. 96–97.

23. "Et ideo invenitur aliquis primus actus absque omni potentia; nunquam tamen invenitur in rerum natura potentia quae non sit perfecta per aliquem actum, et propter hoc semper in materia prima est aliqua forma." *Tractatus de spiritualibus creaturis* 1 resp., ed. Leo W. Keeler (Rome, 1937, repr. 1959), 10.

> subject to quantity. Now it is manifest that quantity is in substance before sensible qualities are. Hence quantities, such as number, dimension, and figures, which are the terminations of quantity, can be considered apart from sensible qualities, and this is to abstract them from sensible matter. But they cannot be considered without understanding the substance which is subject to the quantity, for that would be to abstract them from common intelligible matter. Yet they can be considered apart from this or that substance, and this is to abstract them from individual intelligible matter.
>
> But some things can be abstracted even from common intelligible matter, such as *being* [*ens*], *unity, potency, act,* and the like, all of which can exist without matter, as is plain regarding immaterial things. And because Plato failed to consider the twofold kind of abstraction, as explained above [*ad 1m*], he held that all those things which we have stated to be abstracted by the intellect are abstract in reality.[24]

This looks complicated, but in fact it is simple. Corporeal matter is all that which first has the form of quantity, then the sensible forms of the elements along with their elementary qualities. All that which has actual existence in the material world is determined by form. On account of its essential receptibility, material being can receive an infinite succession of forms; *qua* material, its very nature is to be potency to them.

The doctrine is highly abstract. Its main difficulty consists in identifying matter with being without making it to be a thing. As was to be expected, some Schoolmen insisted that matter was a thing. At first sight this looks like a good idea. Instead of labouring in order to conceive, in Augustine's own terms, a nothing/something, or a non-existing thing, why not conceive

24. *Summa theologiae* 1.85.1 ad 2. Pegis translation. [*Basic Writings of Saint Thomas Aquinas*, ed. Pegis, 1: 815.]

matter as that of which mutability is the very being? One would then call "material" all beings inasmuch as they share in the essence of materiality. Scotus had no doubt that matter was a real being. It is created, and, since it can be the object of an act of creation, it must be a reality distinct from form and "something positive." Since matter has its own entity, it has its own existence, so it is a being in its own right.[25]

The position of Scotus illustrates the misunderstanding which obscures in many minds the true meaning of the peripatetic notion of matter. Because they rightly feel that it cannot possibly be nothing, they conclude that it must be a thing endowed with a being of its own. That is to lose sight of the true notion of being, especially of that kind of being that is coming to be and passing away. That which is becoming is wholly being. True, it is becoming because it is not pure form and pure act; it is both form and matter, act and potency, but that does not mean that such being is composed of being and non-being. Matter itself is being; it is being in potency. In other words, it is being endowed, in virtue of its own nature, that is, from the very fact that it is the kind of a being that it is, with the possibility to become something else. The absence of potentiality is something negative, but the presence of potentiality is something very positive. To have the makings of a champion is not to be a champion, but it is to be a champion in potency, that is, it is to have in itself what it takes to be one. The same remark applies to a possible future scholar, scientist, artist, or mechanic. It is uncharitably said of certain persons that they have no brains, but it can be said of others that they have no hands. To have intellectual or manual capacities is certainly nothing more than to be gifted with certain possibilities, but those possibilities no less certainly are something positive. Their being is but that of the subject to which they belong. They are its matter, and

25. [For Scotus' notion of matter, see Etienne Gilson, *Jean Duns Scot: Introduction à ses positions fondamentales* (Paris, 1952), 432–444.]

by the same token its being is their being. They are its very being in potency to develop, or to be developed into something else.

How that notion got lost, and why it still remains lost to many minds, especially among those who mistake science for philosophy, will be the subject of our third and last lecture.

3. *Matter for Science*

Let us suppose that Thomas' definition of matter is accepted as its philosophical definition: *materia est ens in potentia.*[1] What can it mean to the mind of a modern scientist? The answer is: nothing at all. Hence his conclusion that, since it means nothing to him, it is entirely meaningless. In fact, it is devoid of scientific meaning. The question remains, however: is it devoid of philosophical meaning?

The answer depends on whether one considers the nature of philosophical problems the same or different [from those of science]. Thomas Aquinas would not have understood the sense of the question. I think he would understand it today, just as we do, because he would find himself confronted, as we are, with the grandiose spectacle of modern science, which asks questions about nature foreign to those asked by philosophy, and resorts, in order to answer those questions, to equally unphilosophical methods. Confronted with the physics of Albert Einstein and Max Planck, Aquinas would certainly say: all that is marvelous, and it is true; we Christian theologians should hasten to learn it. And Thomas would probably add: yes, Einstein, Fermi, Planck, Bohr and all those modern physicists are teaching us about the properties of matter – an enormous number of new things we did not know; but they tell us nothing at all about what matter itself is. Now that precisely was our own question. It was a philosophical question, and our answer to it was a philosophical answer. Today, they know much more about science than we did, but they seem to have forgotten the philosophy we used to know. We can now see those things in proper perspective. Why not preserve the two orders of questions and answers? There are no scientific answers to philosophical questions, and vice versa. How are we to distinguish and define those two orders of problems?

1. [See above, p. 91 and p. 102.]

History can help clarify this point. The time when both philosophy, then a very old thing, and science, in the sense of the physico-mathematical interpretation of perceived reality, became hopelessly entangled and confused, can be dated to the first third of the seventeenth century. The name of Descartes can be quoted as a symbol of that event, which was a truly epoch-making one, since it marked the transition from the age of traditional philosophy to that of modern science.

In a letter to Plempius, 3 October 1637, speaking of his physics, Descartes remarked: "Since I am considering nothing else than quantities, figures, and movements, after the manner of the mathematicians, I have closed to myself all the escapes familiar to philosophers."[2] Obviously this program exactly fits the requirements of modern science. There is nothing more in modern physics than sizes, figures, motions, and their mathematical treatment. As for the escapes familiar to the philosophers, and to which Descartes has forbidden himself to resort, what are they if not precisely the properly philosophical notions familiar to Aristotle and the Schoolmen? Descartes was substituting the science of physics for the philosophical physics of his predecessors. In his own language, "true physics" and "mechanics" are synonymous.

What are those "subterfuges" to which philosophers used to resort, and on which Descartes has commented for himself: positively forbidden? Descartes listed some of them in his *Epistola ad G. Voetium*, published in Amsterdam in 1641 in both Latin and Flemish: "Who has ever found anything useful in prime matter, in substantial forms, in occult qualities, and other things of the same kind?"[3] These notions are familiar to historians acquainted with the

2. [*Oeuvres de Descartes*, ed. Charles Adam and Paul Tannery (Paris, 1897–1913), 1: 410–411.]
3. [Descartes' "letter" to Voetius, published as *Principia philosophiae*, appears in ibid., 8.2: 26.]

history of medieval philosophy. We do not hesitate about what they are, namely *philosophical* notions. Since they had no idea of what Descartes and the moderns were going to call science (that is, physico-mathematics) the Schoolmen were not aware of trespassing on scientific ground in developing these philosophical notions. Inversely, Descartes never thought that he was adding physics as a science to physics as a philosophy of nature. To him, his own "true physics" (that is, scientifically true physics) was destined to replace the old philosophical discipline everywhere taught in the schools of his time. He never imagined that the two physics could be true, each in its own order, at one and the same time.

The triumph of scientific physics has therefore been the downfall of the traditional philosophical physics of the schools. And from that defeat there is no appeal, except on one point. In eliminating from minds the philosophical physics of antiquity, scientific physics has simply left all the philosophical problems related to nature not only unanswered but also unasked. In other words, physical science has accumulated an immense amount of knowledge *about* matter, but it has left untouched the old problem of the very nature of matter. How could such a genius as Descartes have not noticed such an oversight? Because his own mind was completely immersed in the discovery of a new universal type of science, uniting the whole body of human knowledge by the common bond of the mathematical method. There would be no place in that science for mere conjectures and probabilities; its whole body would consist of mathematical certitudes. The dawn of that grandiose project can be dated from the year 1620, when Descartes wrote in his notebook the following entry: "I have begun to understand the foundation of an admirable discovery ... ."[4] It may have been the discovery of analytical geometry, or

4. [In his youthful notebook Descartes recorded his discovery of the foundation of a wonderful science ("mirabilis Scientiae fundamenta") in

perhaps already that of his universal mathematics. In any case, the discovery implied the extension of the mathematical method to fields of knowledge hitherto reserved to the abstract dialectics of the philosophers.

This was a colossal gamble. The *Organon* of Aristotle had been conceived as a description of all the methods applicable to the study of all possible objects. It was an axiom to him that every kind of object requires a particular treatment. Nothing is more Cartesian than Spinoza's project of an *Ethica geometrico more demonstrata*;[5] at the same time nothing could possibly be less Aristotelian. Are moral acts of the same nature as geometrical figures, Aristotle would ask? Obviously they are not. Then they should not be handled according to the same method. Circumstances play no part in geometry; to Schoolman Thomas Aquinas nothing is more important in moral acts. They are, Thomas says, the individuating accidents of each human act, its particular conditions which make it to be the very act it is.[6] It is not good to be too good. So whereas what is true of one circle is true of all, the sound judgment of a moral act, for instance, an act of force, temperance, or charity, is not necessarily true of all acts belonging to the same virtue. From the point of view of traditional philosophy, the notion of a geometrically demonstrated ethics is simply preposterous. If you happen to be in a fix, don't apply to a geometer for expert advice; that would be courting disaster.

November 1619. In the same month in 1620, he says that he began to understand the foundation of a marvelous invention: "coepi intelligere fundamentum inventi mirabilis." Gilson discusses these Cartesian musings in René Descartes, *Discours de la méthode*, ed. Etienne Gilson, 3rd ed. (Paris, 1962), 159.]

5. Spinoza, *The Ethics* [*Ethica ordine geometrico demonstrata*], in *The Chief Works of Benedict de Spinoza*, trans. R.H.M. Elwes (New York, 1951), 2: 43–271.

6. *Summa theologiae* 1-2.7.1; 18.10.

Descartes himself never went as far as Spinoza. His ethics is a kind of sensible Stoicism in which, as far as I can see, no trace of geometrism can be detected. But he reformed the whole body of physics, including plant and animal physiology, so as to make it fit for geometrical treatment. The first notion to be reformed was that of matter, which Descartes identified with that of geometrical extension in space according to three dimensions. His list of the properties of matter thus conceived is not always the same. In the third *Meditation* I find this short one: "extension, figure, situation, and motion."[7] In fact, the last three are but modalities of extension: figure is its limits; situation is the relation of its situation in space to that of other figures; motion is reducible to the succession of the positions a moving body occupies in space. A material universe in which there is nothing else is susceptible of an exhaustively geometrical description.

The most visible mark of such a knowledge of physical reality is its practical usefulness. Descartes was right in stressing that point. The physics of Aristotle and the Schoolmen was of no practical value. On the contrary, Descartes was correct in saying that the practical fruitfulness of his own physics was a proof of its truth. The proof of the recipe is in its working. That practical success was the only proof of the scientific truth of the theory. For his experiments to succeed, the structure of matter must be what the scientist says it is, whereas that same success means nothing as to the philosophical truth or untruth of the theory, because philosophical truth is no object of experimental justification. Even if it is true that matter is being in potency, that information about matter is practically useless. On the ground of practical fecundity, philosophy is not in the same league as science; in fact, philosophy cannot even begin to compete with it. Descartes once wrote to a friend that he could not help laughing when he read the scholastic

7. ["De Deo, quòd existat," in] *Oeuvres de Descartes*, ed. Charles Adam and Paul Tannery (Paris, 1897–1913), 7: 43.

definition of motion: the act of what is in potency inasmuch as it is only in potency. What can a physicist or engineer do with that?[8] Remembering what he knows about the laws of motion and their applications in mechanics, the scientist is indeed excusable for laughing, but he should not think he has answered a question he has neglected to ask, namely, what is motion? or what is matter? The scientist alone and not the philosopher will enable us to become, in Descartes' own words, "the masters and possessors of nature."[9] It must be granted that this is the ultimate result of a scientific knowledge of nature. To know if that is its proper object and goal would be another question. As to philosophy, especially metaphysics and ontology, they have no practical purpose; they do not even have the means to pursue one. There is no money in philosophy; even its teachers are not paid to practice philosophy but to teach it.

There is no peril in the error of the philosophers who mistake their philosophy for a science; at worst, they make themselves ridiculous. On the contrary, there is peril in scientism, which consists in mistaking science for a philosophy. Descartes did nothing else on this point, and he did it quite naturally, for just as the Schoolmen had mistaken their philosophy for a science (that is, their philosophical physics for our modern scientific physics), so also Descartes thought that his scientific physics was simply replacing the scholastic philosophy of nature. The result was that, because the philosophical problem of matter does not arise in physico-mathematics, Descartes imagined he had solved it because he had not asked it.

8. [The specific passage has not been identified. Descartes makes a similar point, in less mocking fashion, in his essay *Le monde* (*Oeuvres de Descartes*, ed. Adam and Tannery, 11: 39).]

9. [*Discours de la méthode,* in *Oeuvres de Descartes*, ed. Adam and Tannery, 6: 62; cf. Gilson's edition of the *Discours*, 62, lines 7–8.]

His repeated assertions that matter is only extension and that figure and motion are just modifications of extension in space show that he was mistaking matter for what the Schoolmen had called intelligible matter. Now they called that matter *intelligible* because it was informed with the intelligible form of quantity. From the point of view of scholasticism, there is no matter at all in the physics of Descartes; there is only one single form: extension in space along with its various modalities. The new science was proving its ability to do without a philosophical notion of matter, but the problem of explaining the succession of the various figures in one and the same extension remained unsolved. If matter was extension, then extension would be in potency to its possible configurations and successive positions in space. Descartes could well say that, even if true, the remark is of no scientific interest; but the question remains legitimate in the view of the philosopher, if not in that of the scientist.

Nonetheless, it is not devoid of interest even to the scientist. Leibniz, the great successor of Descartes, was quick to stress that, even to physics, understood as a mathematical science of extended matter, matter could not possibly be reduced to mere extension. First, it must be something extended, and what is that something? Second, a world made up of variously shaped atoms in motion does not find in extension alone the cause of their motion. Some other notion than that of extension, for instance that of *force* or *energy*, is required to explain the fact that atoms or molecules continue to move after being first set in motion by God.

In his 1686 remarks "On the Principles of Descartes," Leibniz has listed many properties of matter that cannot be accounted for by extension alone:

> That extension constitutes the common nature of corporeal substance, I often hear it affirmed with great assurance; affirmed, yes, but proven, never. It is certain that from that extension there follow neither motion nor active force; neither

> from its motion alone do there follow the laws of nature that rule the motions of bodies, as well as their shocks, as I have shown elsewhere. And even the notion of extension is not primitive, but resolvable into other ones. For extension requires one continuous whole in which are to be found several simultaneous coexistences. In short, for there to be extension, since its notion is relative, there must needs be something *extended* or *continued*. As whiteness is in milk, there must be in the body that very thing that constitutes its essence. It is the repetition of that thing, whatever it may be, that constitutes extension.[10]

These objections to Descartes were unanswerable. Leibniz has repeatedly stressed the fact that Cartesian laws of motion were unsatisfactory for the very reason that he had reduced matter to mere extension without taking force into account. In chapter XVII of the *Discourse on Metaphysics*, Leibniz has demonstrated, against Descartes and some Cartesians, that God is not always preserving the same quantity of movement in the universe but the same quantity of force. Now force and quantity of movement are very different. Force must be calculated by the quantity of the effect it can produce, for example by the height to which a certain body, of a certain weight and size, must be raised in order to obtain that effect. Now that is very different from the speed one can give that body. It takes more than double the force to get double the speed.[11]

This is one more remarkable instance of what I have (quite unsuccessfully) proposed to call metaphysical experiment in the

10. [Leibniz, "Critical Thoughts on the General Part of the Principles of Descartes," in his *Philosophical Papers and Letters*, ed. and trans. L.E. Loemker (Chicago, 1956), 2: 641–642. Gilson was using another (or his own) translation.]
11. Leibniz, *Discourse on Metaphysics* 17, trans. George Montgomery, 2nd ed. (Chicago; London, 1918), 29–32.

history of philosophy.[12] For Descartes had refuted the philosophical physics of the Schoolmen and their leader Aristotle by first establishing that there are no forms in matter but only extension. Where there is only extension there can be mechanical motion, as was the case in Descartes' molecular physics, but no force. But experience shows that "there is a great difference between quantity of movement and force."[13] To which Leibniz immediately adds the decisive remark that the force in question is something different from the size and figure of the moving body and from its movement. So bodies cannot be conceived as consisting of only extension and its modifications

> as our moderns persuade themselves. We are therefore obliged to restore certain beings or forms which they have banished. It appears more and more clear that although all the particular phenomena of nature can be explained mathematically or mechanically by those who understand them, yet nevertheless, the general principles of corporeal nature and even of mechanics *are metaphysical rather than geometrical* and have reference to certain indivisible forms or natures as the causes of the appearances, rather than to the corporeal mass or to extension.[14]

The next chapter of the *Discourse* is devoted to a vindication, against Descartes, of the notion of final causes.[15] The doctrine of Leibniz, as well as that of his disciple Christian Wolff,[16] seems to mark the point of perfect equilibrium between the philosophy without the physico-mathematics of Aristotle and the Schoolmen [on the one hand] and the physico-mathematics without meta-

12. [See Gilson, *The Unity of Philosophical Experience* (New York, 1937).]
13. *Discourse on Metaphysics* 17, trans. Montgomery, 31.
14. Ibid., 18, trans. Montgomery, 32–33 [Gilson's emphasis].
15. Ibid., 19, trans. Montgomery, 33–35.
16. Christian Wolff [*Philosophia prima sive ontologia*, ed. Joannes Ecole (Hildesheim, 1962)].

physics of Descartes and the moderns [on the other]. But it is a fact that, though he was wrong, Descartes carried the day.

The result of Descartes' victory has been that modern scientism leaves us without any metaphysical notion of matter, which can hardly be counted as a progress. Of course, science has discovered an unbelievable quantity of truth concerning the structure of material bodies, for instance practically the whole of chemistry. It has pushed beyond the notion of atoms and discovered in elementary molecules an infinitely small universe of unbelievable complexity. The molecules of elements consist of one or more similar atoms. Atoms themselves consist of "a complex arrangement of electrons revolving about a positively charged nucleus containing [except for hydrogen] protons and neutrons,"[17] plus an undetermined and, as it seems, unlimited number of other particles, of which the proliferation in modern physics is not its least disturbing aspect. The practical results obtained by atomic and subatomic science are so conspicuous that the absence of a philosophical notion of matter is hardly noticeable. In fact, since modern science does not even reduce it to extension, as Descartes had done, it does very well without such a notion, as long at least as it does not entertain the illusion of having one. That the harm caused by that illusion is merely philosophical explains why it goes unnoticed by most scientists, but we philosophers cannot consider it unimportant.

If a philosopher considers the attitude of modern physicists with respect to the question, what he first notices is that they are so thoroughly unaware of its meaning that they are not even aware of not asking it. Because they are busy investigating the nature of matter, they believe they are examining the question of what matter is. They have lost Leibniz' awareness of the philosophical meaning of the question. Deaf to the good advice of

17. *Webster's New World Dictionary of the American Language*, ed. David B. Guralnik, 2nd college ed. (New York, 1970), 88.

Leibniz, they have neglected to re-read Plato in order to recapture their lost feeling for the presence of intelligible elements in the structure of material reality.[18]

I have stressed the case of the conflict between Leibniz and Descartes because it helps us to situate the precise point on which the conflict between Aristotelian physics and that of the moderns has arisen. It has been said by an excellent scholar, the late professor William Carlo of the University of Ottawa, that the main point of disagreement between the philosophy of the Schoolmen and modern philosophers is not about metaphysics but about physics.[19]

I think my friend Professor Carlo was much more competent than I am on that subject. So I imagine that I am missing his point in disagreeing with him on the question; but I feel bound to voice my disagreement, because it bears on the precise point I am trying to make. I would not say that the philosophy of the Schoolmen disagrees with modern philosophy about physics rather than about metaphysics. I would rather locate the disagreement in the fact that the Schoolmen had no physics in the modern sense of physico-mathematics,[20] while moderns have no metaphysics at all.

It has taken a long time to achieve a clear-cut distinction between philosophical physics and scientific physics. The distinction between mathematics and philosophy has always been more or less confusedly present to the minds of the Schoolmen. They

18. Leibniz, *Discourse on Metaphysics* 35–36.

19. [William Carlo, "Biological Emergence and the Plurality of Sciences: An Alternative to Mach," in International Congress of Philosophy, *Akten des XIV Internationalen Kongresses für Philosophie: Wien, 2–9 September, 1968* (Vienna, 1968–1971), 4: 424–32. I am indebted to Fr. Lawrence Dewan, OP, for this reference.]

20. [This general statement should be qualified by Gilson's own words about the beginning of mathematical physics (for example, in Roger Bacon) in the Middle Ages. See his *History of Christian Philosophy in the Middle Ages* (New York, 1955), 309–310.]

could handle any philosophical discipline by means of logic alone, but they were painfully aware that mathematics, geometry, and all the sciences dealing with quantity, either discontinuous or continuous, prerequired the mastery of intellectual techniques to which the *Organon* of Aristotle provided no opening. The *Principles of Philosophy* of Descartes covers the whole field from metaphysics to mechanics, geology, chemistry, and physics.[21] All these disciplines constitute one single perfectly homogeneous body of knowledge. As for Leibniz, it is in his *Discourse on Metaphysics*, and on the strength of metaphysical principles, that he establishes, against those of Descartes, the true laws of motion.

In the twilight of the seventeenth century, in the preface to the first edition of his *Mathematical Principles of Natural Philosophy* (8 May 1686), Newton also began by taking side with the moderns, "rejecting substantial forms and occult qualities," and by offering his work "as the mathematical principles of philosophy"; for the whole burden of philosophy consists in this: "from the phenomena of motions to investigate the forces of nature, and then from those forces to demonstrate the other phenomena."[22] Newton's impression was not that he was adding to ancient philosophy an entirely new science but that he was substituting a new and true philosophy of nature for the ancient and false one.

That unitary conception of human knowledge has found its belated and truly genial exponent in the person of Auguste Comte, the modern Schoolman, who proceeded in his *Course of Positive Philosophy*[23] to a complete exposition of the sciences of his

21. *Philosophical Writings*, ed. and trans. Elizabeth Anscombe and Peter Thomas Geach (Edinburgh, 1954), 183–238.
22. Isaac Newton, *Philosophiae naturalis principia mathematica* [ed. Alexandre Koyré and I. Bernard Cohen (Cambridge, Mass., 1972), 1: 15–16. This edition is based on the 3rd ed. (London, 1726), and contains the Preface to the first edition.]
23. [Auguste Comte, *Cours de philosophie positive* (Paris, 1908–1934) and *A General View of Positivism*, trans. J.H. Bridges (New York, 1957). Gilson comments at

time, at least as to their methods and principles, and finally undertook to create a still missing positive science, namely sociology. Comte could not fail to encounter the general notion of matter, and his personal conception of the hierarchical structure of reality provided him with an answer. Comte taxed with materialism all attempts to reduce the superior to the inferior: physics to mathematics, biology to physics, sociology to biology. In his view, every order of phenomena, both more general and simple, was material with respect to less general and more complex orders of reality. There was something sound in that conception of materiality, which resembled the Aristotelian identification of matter with determinability and potentiality, but Comte was the last philosopher to attempt an all-embracing interpretation of reality, including the whole encyclopedia of the sciences.[24]

I am not qualified to speak to you about the modern science of matter; what little I know about it leaves me bewildered. The number of elementary particles discovered by the physicists is amazing. Especially surprising to my ignorant mind is their remarkable habit of going by pairs of opposites, so much so that when a scientist discovers a new kind of subatomic particle, he can safely predict the discovery of another one exactly contrary to it. After the discovery of the proton, the history of the discovery of the neutron, as told by Donald J. Hughes,[25] reads like a fascinating novel. But there are countless other physical entities, such as neu-

---

length on Comte in his *Unity of Philosophical Experience* (New York, 1937; many reprints), 248–270. See also Etienne Gilson, Thomas Langan, and Armand Maurer, *Recent Philosophy: Hegel to the Present* (New York, 1966), 267–276.]

24. [Gilson added a note here, seemingly unconnected to what precedes it: "Today, matter as a philosophical problem has ceased to exist. One has never known so much about matter nor been so little interested in the philosophical nature of materiality."]

25. *The Neutron Story* (Garden City, N.Y., 1959).

trinos, photons, waves and/or molecules, and the quanta of energy, including the mysterious Planck constant that defines the relation between the frequency of the radiation and the dimension of the quantum. It is designated by the letter h (h = 6.6×10^{-27} erg second). Nobody knows the reason for that constant, but it works. The photons are quanta of light which, by the way, is a material body, not a spiritual reality, as was believed by many Schoolmen. As [Sir Arthur] Eddington has put it, in a short poem inspired by Einstein's theory and its experimental verification by Michelson:

> One thing at least is certain,
> LIGHT HAS WEIGHT. ...
> Light rays when near the Sun,
> DO NOT GO STRAIGHT.[26]

Man has never known as much about matter as he now does, but never has matter been less conceived as a formless something, *nec quid, nec quale, nec quantum*. Like the traditional matter of philosophy, it is the seat and subject of becoming or change, but never has being in potency been laden with such a wealth of actual determinations. In the language of modern physicists, to have weight is to have matter, and vice versa. Mass is a certain quantity of matter, but matter itself and its mass is convertible with energy, according to Einstein's celebrated formula: $E = mc^2$ (energy equals mass multiplied by the square of the speed of light). The enormity of the energy released by the dissociation of matter is owing to the size of c, the speed of light, which is 3×10^{10} [cm/sec] (three multiplied by 10 power 10 [cm/sec], that is, about two billion kilowatt hours).[27] What becomes of matter in that physics? It certainly is nothing inert like the geometrical extension

26. [In Ronald W. Clark, *Einstein: The Life and Times* (New York, 1971), 229.]
27. [*Sic* in text. One kilogram of mass could potentially produce about 25 billion kilowatt hours of energy.]

of Descartes. On the contrary, it seems to be a kind of concretized energy, or energies, that stand on the side of form.

This clearly appears in what physicists call "antimatter." Antimatter is still matter, but it is a matter consisting of anti-particles. An antiparticle is a particle symmetrical with an elementary particle and able to annihilate itself along with it, thereby liberating energy under the form of photons (that is, quanta of energy, the flux of which constitutes electro-magnetic radiation). For instance, the antiproton is the antiparticle of the proton, and so also the antineutron, etc. Nothing is less indeterminate than modern matter, since it arises from energy and can dissolve into it again. In other words, matter is now a purely physical and scientific notion, on which the philosopher as such has nothing to say, for or against, while, on the contrary, the philosophical notion of matter conceived as the subject of the very possibility of change, motion, or becoming in general, has become entirely foreign to modern scientific minds. And rightly so, for the science of physics presupposes the givenness of its own object. The only science from which physics can receive its object in our own time, as in those of Thomas Aquinas or Aristotle, is metaphysics.

A scientist as such is therefore justified in attributing no scientific value to the philosophical definition of matter as being in potency, but that is no justification for considering it as meaningless. First of all, despite all their efforts, the philosophers have failed to improve upon it. Secondly, those among the great scientists who are gifted for philosophical reflection are acutely aware of a truth naturally familiar to poets and other artists, namely, the gap between nature and our scientific knowledge of it.

By that I do not simply mean that there are more things in heaven and earth than science can dream of; that would be nothing. The point is that the nature of reality is other than that of our knowledge of it. More plainly, the intellectual transcription of natural reality is essentially different from that reality. Not knowing how to convey the feeling of that difference, I shall resort to two anecdotes, one personal, the other involving, on the contrary, a witness of world fame.

I once read with passionate interest the *Journal* of the painter Eugène Delacroix.[28] I remember feeling amused by a trait that looked to me a mark of puerility in such a great artist. Delacroix hated naturalists, especially zoologists. I could not understand why. Later in life, as I happened to be in South America – to be exact, in Colombia – visiting the zoological collections of the University of Santa Fé in Bogotá, I expressed to my guide my admiration for a beautiful bird the size of the common fat blackbird in that country. My learned colleague and guide told me that the plumage of those birds varied from province to province; but, he added, we have specimens of all those varieties right here. Whereupon he unexpectedly opened a drawer just in front of me, and it was full of desiccated little corpses of those warbling beauties. They were all there, dead. A violent feeling of revulsion invaded me, with, yes, an anger against the authors of that desecration of life and natural beauty. I then remembered Delacroix, and for the first time in my personal experience I shared in what my reason still cannot help considering that childish feeling of his. What I was then seeing was science turning its back on reality.

In Ronald Clark's fascinating biography of Einstein, *Einstein: The Life and Times*, the wife of the physicist Max Born relates that one day she asked Einstein: "Do you believe that absolutely everything can be expressed scientifically?" "Yes," he replied, "it would be possible, but it would make no sense. It would be description without meaning – as if you described a Beethoven symphony as a variation of wave pressure." To which Mrs Max Born simply adds

28. *Journal de Eugène Delacroix*, ed. André Joubin (Paris, 1932; rev. ed. 1950).

the remark, "This was a great solace to me."[29] Einstein was very fond of music and liked to play the violin. An exhaustive acoustic analysis of Mozart's divine *Ave verum corpus* would have given Einstein complete intellectual satisfaction; but it would have been science, not music.

One can think of a third attitude with respect to music: neither simply play it or enjoy it, nor analyze it into its component sounds, but wonder what it is. What would Einstein have missed if he had been left with acoustics and the physical description of Beethoven's music? Simply music itself. Not so with a philosophical knowledge of music. The sense experience of reality is the basis and origin of all our knowledge of it. Practical knowledge rests upon it; philosophical knowledge is its immediate interpretation in terms of concepts; scientific knowledge is its translation in terms of extension and number. But the relation of science and philosophy to sense experience is not the same. Philosophy accepts it such as it is given to sense, including quality as well as quantity. Hence the consequence that philosophy has for its object reality itself and not, as is the case with science, nature stripped of quality and reduced to quantity.

The philosophy of Aristotle conquered the schools of the Middle Ages precisely because it was always about reality as perceived by the senses and translated into terms of conceptual knowledge: being, cause, becoming, quantity, quality, relation, and so on. The peculiar kind of satisfaction which the mind derives from it is due to the same reason that makes it unsatisfactory to the scientists. Because it starts from reality such as it is given to the senses, we always know what we are talking about, but its concepts of reality suffer the same obscurity as reality itself. Consider being, for instance. We all know a being when we see one, but to say what being is, is quite a different

29. *Einstein: The Life and Times* [(New York, 1971)], 192.

proposition. So also with becoming, cause, form, or matter. That matter is being in potency is both evident and obscure, because we all see the potentiality of what is under our very eyes, becoming what it was able to become, and was not. Matter is only one among those philosophical translations of sense experience. Science is not interested in it; philosophy fails to turn it into a clear and distinct notion of the mind. And yet philosophy prefers to maintain it such as it is, because to suppress it is to mutilate reality by removing one of its essential parts.

E.G., Cravant (Yonne), 30 July 1972

Appendix

[At the end of his projected lectures Gilson added these additional notes on four philosophers who dealt with the concept of matter.]

1. Plato

The empirical nature of the notion of matter appears from the fact that, before Aristotle, Plato already received it as a given fact, although it is "difficult of explanation and dimly seen" (*Timaeus* 49 [in *The Dialogues of Plato,* trans. Benjamin Jowett, 3rd ed. (New York, 1937), 2: 29]). Plato goes on to say: "What nature are we to attribute to this new kind of being? We reply that it is the receptacle, and in a manner the nurse, of all generations" (ibid.). Supposing there are different objects made of gold, and we are asked: "What is it?" the safest answer is to say: "gold." "And the same argument applies to the universal nature which receives all bodies – that must be always called the same; for, while receiving all things, she never departs at all from her own nature, and never in any way, or at any time assumes a form like that of any of the things which enter into her" (ibid., 50 [in *The Dialogues of Plato,* trans. Jowett, 2: 30]). – "Wherefore, the mother and receptacle of all created and visible and in any way sensible things is not to be termed earth, or air, or fire, or water, or any of their compounds or any of the elements from which these are derived, but is an invisible and formless being which receives all things and in some mysterious way partakes of the intelligible, and is most incomprehensible" (ibid., 51 [in *The Dialogues of Plato,* trans. Jowett, 2: 31]).

In less technical terms, this passage contains the whole Aristotelian notion of matter, in the state of pure intuition.

2. Plotinus

In Plotinus, matter essentially remains what it was in Plato, with a touch of mystical feeling proper to him.

"We utterly eliminate every kind of Form; and the object in which there is none whatever, we call Matter: if we are to see Matter we must so completely abolish Form that we take shapelessness into our very selves" (*Enneads* 1.8.9, [in *The Six Enneads*, trans. Stephen MacKenna and B.S. Page, Great Books of the Western World, ed. Robert M. Hutchins (Chicago, 1952)] 17: 31). Matter is evil in the sense of not having quality (ibid., 10, [trans. MacKenna and Page, Great Books] 17: 32). This view, foreign to Aristotle and his medieval disciples, suggests that, in the mind of Plotinus, matter is enough of a thing to be susceptible of qualification. "No one says that matter has no nature ..." (Ibid.). On the way the intellect takes shapelessness into itself (ibid., 2.4.10, [trans. MacKenna and Page, Great Books] 17: 53). Perhaps the most impressive passage on matter is ibid., 3.6.7 [trans. MacKenna and Page, Great Books 17: 111]: "Matter must be bodiless .... Matter is no soul ... it lives on the farther side of all these categories and so has no title to the name of Being. It will be more plausibly called a non-being, ... so that it is no more than the image and phantasm of mass, a bare aspiration towards substantial evidence; it is stationary but not in the sense of having position, it is in itself invisible, eluding all effort to observe it, present where no one can look, ... a phantasm unabiding and yet unable to withdraw - not even strong enough to withdraw, so utterly has it failed to accept strength from the Intellectual Principle, so absolute its lack of all Being," etc.

3. St Augustine

"And still that which I conceived was without form, not as being deprived of all form, but in comparison of more beautiful forms; and true reason did persuade me that I must utterly uncase it of all remnants of form whatsoever, if I would conceive matter absolutely without form; and I could not; for sooner could I imagine that not to be at all, which should be deprived of all form, than conceive a thing between form and nothing, neither formed nor nothing. So my mind gave over to question thereupon with my spirit, it being filled with the images of formed bodies, and changing and varying them as it willed; and I bent myself to the bodies themselves, and looked more deeply into their changeableness, by which they cease to be what they have been, and begin to be what they were not; and this same shifting from form to form, I suspected to be through a certain formless state, not through a mere nothing .... For the changeableness of changeable things is itself capable of all those forms into which these changeable things are changed. And this changeableness, what is it? Is it soul? Is it body? Might one say 'a nothing something,' an 'is, is not' .... I would say, this were it; and yet in some way, was it even then, as capable of receiving these visible and compound figures" (*Confessions* 12.6.6, [in *The Confessions of St Augustine,*] trans. E.B. Pusey [London; New York, 1907, repr. 1909], 280–281.)

4. Denis Diderot

Denis Diderot has often given excellent examples of philosophical oversights mistaken for scientific sights. For instance, in his celebrated *D'Alembert's Dream*, the dreamer is supposed to be talking in his sleep and to say: "A living point .... No, I am wrong. First, nothing, then one living point .... To that living point another one joins itself, then another one, and from those successive applications (of point to point) there results one being that is one, for surely I am one, I cannot doubt it ... ." [*Oeuvres complètes de Diderot*, ed. J. Assézat (Paris, 1875) 2: 124 (Gilson's translation).] The dream discourse is sensible and consistent, only it consists of philosophical unintelligibilities. First, there is nothing, then, to a philosophical mind, next there still will be nothing: *ex nihilo nihil*. In the dream of d'Alembert, on the contrary, from nothingness, or at least after it, there arises one point, which is surprising, and to make it more surprising still, that point is a living point. The philosopher will wonder: granting the appearing of the miraculous point, what makes it to be one living point? What is life? Next comes a further difficulty: how do those successively self-adding points come to form one living composite unit? Not only do all these philosophical problems not exist for Diderot but, to him, to pass them by is the mark of a scientific turn of mind. In his *Entretien entre D'Alembert et Diderot,* he calls philosophical considerations in general "Galimatias métaphysico-théologique" ["metaphysico-theological gibberish"] (ibid., 2: 116).

Bibliography

Adler, Mortimer J. "Problems for Thomists." *The Thomist* 2 (1940): 88–155.

—. *Problems for Thomists: The Problem of Species.* New York: Sheed & Ward, 1940.

—. "Solution of the Problem of Species." *The Thomist* 3 (1941): 279–379.

Albert the Great. *Summa de creaturis.* In *B. Alberti Magni Ratisbonensis episcopi, ordinis Praedicatorum, Opera omnia, ex editione lugdunensi religiose castigate* [*Opera Omnia*], ed. Auguste Borgnet, 34: 307–798. 38 vols. Paris: Vivès, 1890–1895.

Apuleius. *The Golden Ass, being the Metamorphoses of Lucius Apuleius.* Trans. W. Adlington, revised by S. Gaselee. Loeb Classical Library. 1915. Repr. London: Heinemann; New York: G.P Putnam's Sons, 1919.

Aquinas. *See* Thomas Aquinas.

Aristotle. *The Basic Works of Aristotle.* Ed. Richard McKeon. New York: Random House, 1941.

—. *Categoriae.* Trans. E.M. Edghill. In *The Basic Works of Aristotle,* 3–37.

—. *De partibus animalium.* Trans. William Ogle. In *The Basic Works of Aristotle,* 641–661.

—. *Metaphysica.* Trans. W.D. Ross. In *The Basic Works of Aristotle,* 681–926. [For a revised translation, see *Metaphysics,* trans. W.D. Ross, in *The Complete Works of Aristotle: The Revised Oxford Translation,* ed. Jonathan Barnes. 2 vols. Bollingen Series 71.2. Princeton, N.J.: Princeton University Press, 1984.]

—. *Physica.* Trans. R.P. Hardie and R.K. Gaye. In *The Basic Works of Aristotle,* 213–394.

Augustine. *Confessions.* In *The Essential Augustine,* ed. and trans. Vernon J. Bourke. Mentor-Omega Books. New York; Toronto: New American Library, 1964.

—. *The Confessions of St Augustine.* Trans. E.B. Pusey. Everyman's Library. London: Dent; New York: Dutton, 1907, repr. 1909.

—. *Soliloquies: Augustine's Inner Dialogue.* Ed. John E. Rotelle, OSA. Trans. Kim Paffenroth. Augustine Series 2. Hyde Park, N.Y.: New City Press, 2000.

Avicenna. *Liber primus naturalium: tractatus primus de causis et principiis naturalium.* Ed. S. Van Riet. Avicenna latinus 8. Louvain-la-Neuve: Peeters; Leiden: Brill, 1992.

Bacon, Roger. *See* Roger Bacon.

Bergson, Henri. *Les deux sources de la morale et de la religion.* Bibliothèque de philosophie contemporaine. Paris: Alcan, 1932; many reprints. [Translated by R. Ashley Audra and Cloudesley Brereton as *The Two Sources of Morality and Religion.* London: Macmillan, 1935; many reprints.]

—. *L'évolution créatrice.* Paris: Alcan, 1907; many reprints. [Translated by Arthur Mitchell as *Creative Evolution.* New York: Henry Holt, 1911; many reprints.]

Biblia Sacra iuxta Vulgatam versionem. Ed. Robert Weber. Stuttgart: Deutsche Bibelgesellschaft, 1969; 3rd ed. 1983.

Browne, Sir Thomas. *Religio medici.* Ed. James Winny. Cambridge: Cambridge University Press, 1963.

Buffon, Georges-Louis Leclerc. *Histoire naturelle.* Vol. 12 of *Oeuvres complètes de Buffon.* Ed. Bernard-Germain-Etienne de la Ville-sur-Illon, comte de Lacépède. 26 vols. Rev. ed. Paris: Eymery, 1825–1828.

Carlo, William. "Biological Emergence and the Plurality of Sciences: An Alternative to Mach." In International Congress of Philosophy, *Akten des XIV Internationalen Kongresses für Philosophie: Wien, 2–9 September, 1968,* 4: 424–432. 6 vols. Vienna: Herder, 1968–1971.

Cicero. *De natura deorum: Academica.* With an English translation by H. Rackham. Loeb Classical Library. London: Heinemann; New York: G.P. Putnam's Sons, 1933.

Clark, Ronald W. *Einstein: The Life and Times*. New York: World Publishing Co., 1971.

Cohen, Gustave. *Ronsard, sa vie et son oeuvre*. 5th ed. Paris: Gallimard, 1956.

Comte, Auguste. *Cours de philosophie positive*, 6 vols. Paris: Ancienne Librairie Schlecher, Alfred Costes, 1908–1934.

—. *A General View of Positivism*. Trans. J.H. Bridges. New York: Robert Speller and Sons, 1957. [Official centenary edition of the International Auguste Comte Centenary Committee.]

Cuénot, Lucien. *L'espèce*. Encyclopédie scientifique. Bibliothèque de biologie générale. Paris: Doin, 1936.

—. *L'évolution biologique; les faits, les incertitudes*. Paris: Masson, 1951.

Cuvier, Georges. *Le règne animal distribué d'après son organisation, pour servir de base à l'histoire naturelle des animaux et d'introduction à l'anatomie comparée*. 2nd ed. 11 vols. in 17. Paris: Fortin, Masson, 1836–1849.

Darwin, Charles. *The Autobiography of Charles Darwin, 1809–1882: With Original Omissions Restored*. Ed. Nora Barlow. London: Collins, 1958.

—. *The Origin of Species by Means of Natural Selection*. 6th ed. London: Murray, 1872; many reprints.

—. *The Origin of Species by Means of Natural Selection; The Descent of Man and Selection in Relation to Sex*. Great Books of the Western World, ed. Robert M. Hutchins, 49. Chicago: Encyclopaedia Britannica, 1952.

Deely, John N. "The Philosophical Dimensions of *The Origin of Species*." *The Thomist* 33 (1969): 75–149, 251–335.

Delacroix, Eugène. *Journal de Eugène Delacroix*. Ed. André Joubin. 3 vols. Paris: Plon, 1932; rev. ed. 1950.

Descartes, René. *Discours de la méthode*. Ed. Etienne Gilson. 3rd ed. Bibliothèque des textes philosophiques. Paris: Vrin, 1962; repr. 1964.

—. *Oeuvres de Descartes*. Ed. Charles Adam and Paul Tannery. 11 vols. in 13. Paris: Cerf, 1897–1913; repr. Paris: Vrin, 1964–1974.

—. *Philosophical Writings*, ed. and trans. Elizabeth Anscombe and Peter Thomas Geach. Nelson Philosophical Texts. Edinburgh: Nelson, 1954.

Diderot, Denis. *Oeuvres complètes de Diderot*. Ed. J. Assézat. 20 vols. Paris: Garnier Frères, 1875–1877.

Dobzhansky, Theodosius. *Genetics and the Origin of Species*. Columbia Biological Series 11. New York: Columbia University Press, 1937; 3rd ed. 1951, repr. 1964.

—. *Heredity and the Nature of Man*. New York: Harcourt, Brace & World, 1964.

Forest, Aimé. *La structure métaphysique du concret selon saint Thomas d'Aquin*. Étude de philosophie médiévale 14. Paris: Vrin, 1931.

Gilson, Etienne. *The Christian Philosophy of St Thomas Aquinas*. Trans. Laurence K. Shook. New York: Random House, 1956; repr. New York: Octagon Books, 1983; repr. Notre Dame, Ind.: University of Notre Dame Press, 1994. [Originally published as *Le thomisme: Introduction à la philosophie de Saint Thomas d'Aquin*. 5th ed. Études de philosophie médiévale 1. Paris: Vrin, 1944.]

—. *From Aristotle to Darwin and Back Again: A Journey in Final Causality, Species, and Evolution*. Trans. John Lyon. Notre Dame, Ind.: University of Notre Dame Press; London: Sheed & Ward, 1984. [Originally published as *D'Aristote à Darwin et retour: Essai sur quelques constantes de la biophilosophie*. Essais d'art et de philosophie. Paris: Vrin, 1971.]

—. *History of Christian Philosophy in the Middle Ages*. Random House Lifetime Library. New York: Random House, 1955.

—. *Jean Duns Scot: Introduction à ses positions fondamentales*. Études de philosophie médiévale 42. Paris: Vrin, 1952.

—. "Mon ami Lévy-Bruhl, philosophe, sociologue, analyste des mentalités primitives." *Nouvelles littéraires*, 18 March 1939, 1.

—. *Le réalisme méthodique.* Cours et documents de philosophie. Paris: Téqui, 1936. [Ch. 3 translated by D.A. Patton as "Concerning Christian Philosophy: The Distinctiveness of the Philosophic Order," in *Philosophy and History: Essays Presented to Ernst Cassirer*, ed. Raymond Klibansky and H.J. Paton (Oxford: Clarendon Press, 1936), 61–76; repr. as "The Distinctiveness of the Philosophic Order" in *A Gilson Reader*, ed. Anton C. Pegis, Doubleday Image Book D55 (Garden City, N.Y.: Image Books, 1957), 49–63. The entire book has been translated by Philip Trower as *Methodical Realism* (Front Royal, Va.: Christendom Press, 1990).]

—. "Réflexions sur l'éducation philosophique." *Conférence* 26 (2008): 611–631.

—. "Le rôle de la philosophie dans l'histoire de la civilisation." In *Proceedings of the Sixth International Congress of Philosophy, Harvard University, ... 1926*, ed. Edgar Sheffield Brightman, 529–535. New York: Longmans, Green, 1927.

—. "Sur le positivisme absolu." *Revue philosophique de la France et de l'étranger* 68 (1909): 63–65.

—. *Thomism: The Philosophy of Thomas Aquinas.* Trans. Laurence K. Shook and Armand Maurer. Etienne Gilson Series 24. Toronto: Pontifical Institute of Mediaeval Studies, 2002. [Originally published as *Le thomisme: Introduction à la philosophie de Saint Thomas d'Aquin.* 6th ed. Études de philosophie médiévale 1. Paris: Vrin, 1965.]

—. *The Unity of Philosophical Experience.* New York: Charles Scribner's Sons, 1937; many reprints.

—, Thomas Langan, and Armand Maurer. *Recent Philosophy: Hegel to the Present.* A History of Philosophy 4. New York: Random House, 1966.

Goethe, Johann Wolfgang von. *Zahme Xenien* [*Gentle Ironies*]. In *Sprüche in Reimen, Zahme Xenien, und Invektiven.* Ed. Max Hecker. Leipzig: Insel, 1908.

Graefrath, Bernd. "Darwinism: Neither Biologistic nor Metaphysical." In *Darwinism and Philosophy*, ed. Vittorio Hösle and Christian Illies, 364–380. Notre Dame, Ind.: University of Notre Dame Press, 2005.

Harper's Latin Dictionary: A New Latin Dictionary Founded on the Translation of Freund's Latin-German Lexicon. Ed. E.A. Andrews, Charlton T. Lewis, and Charles Short. New York: Harper & Brothers; Oxford: Clarendon Press, 1879.

Holmes, Oliver Wendell. *The Autocrat of the Breakfast Table*. Everyman's Library. London: Dent; New York: Dutton, 1906.

Hughes, Donald J. *The Neutron Story*. Science Study Series S1. Garden City, N.Y.: Doubleday Anchor Books, 1959.

James, William. *The Principles of Psychology*. Great Books of the Western World, ed. Robert M. Hutchins, 53. Chicago: Encyclopaedia Britannica, 1952.

—. "A World of Pure Experience." *Journal of Philosophy, Psychology, and Scientific Methods* 1 (1904): 533–543, 561–570.

Kitcher, Philip. "Species." In *The Units of Evolution: Essays on the Nature of Species*, ed. Marc Ereshefsky, 317–342. Cambridge, Mass.: MIT Press, 1992.

La Fontaine, Jean de. *The Fables of La Fontaine*. Trans. Marianne Moore. Compass Books C146. New York: Viking Press, 1964.

Lamarck, Jean-Baptiste. *Lamarck: Pages choisies*. Ed. Lucien Brunelle. Les classiques du peuple. Paris: Éditions sociales, 1957.

—. *Philosophie zoölogique, ou Exposition des considérations relative à l'histoire naturelle des animaux*. 2 vols. Paris: Dentu et l'auteur, 1809.

Leibniz, Gottfried Wilhelm. "Critical Thoughts on the General Part of the Principles of Descartes." In *Philosophical Papers and Letters*, ed. and trans. L.E. Loemker. 2 vols. Chicago: University of Chicago Press, 1956.

—. *Discourse on Metaphysics: Correspondence with Arnauld, and Monadology*. Trans. George Montgomery. Religion of Science Library 52. Chicago; London: Open Court, 1902; 2nd ed. 1918.

—. "The Principles of Nature and of Grace, Based on Reason." In *Philosophical Papers and Letters*, ed. and trans. L.E. Loemker. 2 vols. Chicago edition. Chicago: University of Chicago Press, 1956.

— and Antoine Arnauld. *The Leibniz-Arnauld Correspondence.* Ed. and trans. H.T. Mason. Philosophical Classics. Manchester: Manchester University Press, 1967.

Livy. *Livy* [*Ab urbe condita*]. Trans. B.O. Foster, F.G. Moore, Evan T. Sage, and A.C. Schlesinger, with an index by Richard M. Geer. 14 vols. Loeb Classical Library. London: Heinemann; Cambridge, Mass.: Harvard University Press, 1919–1959.

Macbeth, Norman. "The Species Problem, or the Origin of What?" Ch. 3 in *Darwin Retried: An Appeal to Reason.* Boston: Gambit, 1971.

Maritain, Jacques. "Concerning a Critical Review." *The Thomist* 3 (1941): 45–53.

—. Foreword to Mortimer J. Adler, *Problems for Thomists: The Problem of Species.* New York: Sheed & Ward, 1940.

—. *Philosophy of Nature.* Trans. Imelda C. Byrne. New York: Philosophical Library, 1951.

Maurer, Armand. "Darwin, Thomists, and Secondary Causality." *The Review of Metaphysics* 57 (2004): 491–514.

—. "Etienne Gilson, Critic of Positivism." *The Thomist* 71 (2007): 199–220.

Mayr, Ernst. *This is Biology: The Science of the Living World.* Cambridge, Mass.: Belknap Press of Harvard University Press, 1997.

Molière [Jean-Baptiste Pocquelin]. *Le malade imaginaire.* In *Oeuvres complètes*, ed. Maurice Rat. 2 vols. Paris: Gallimard, 1956.

Murphy, Francesca Aran. *Art and Intellect in the Philosophy of Etienne Gilson.* Eric Voegelin Institute series in political philosophy. Columbia, Mo.: University of Missouri Press, 2004.

Nelson, Ralph. "Two Masters, Two Perspectives: Maritain and Gilson on the Philosophy of Nature." In *Wisdom's Apprentice: Thomistic Essays in Honor of Lawrence Dewan, O.P.*, ed. Peter A.

Kwasniewski, 214–236. Washington, D.C.: Catholic University of America Press, 2007.

Newton, Isaac. *Philosophiae naturalis principia mathematica.* Ed. Alexandre Koyré and I. Bernard Cohen. Cambridge, Mass: Harvard University Press, 1972. [Based on the 3rd ed., London: Innys, 1726.]

Pascal, Blaise. *Pensées,* ed. Michel Le Guern. 2 vols. Collection Folio 936–937. Paris: Gallimard, 1977.

Pasnau, Robert. *Thomas Aquinas on Human Nature: a Philosophical Study of Summa theologiae 1a, 75–89.* Cambridge: Cambridge University Press, 2002.

Plato. *The Dialogues of Plato.* Trans. Benjamin Jowett. 3rd ed. 2 vols. New York: Random House, 1937.

Plotinus. *The Six Enneads.* Trans. Stephen MacKenna and B.S. Page. Great Books of the Western World, ed. Robert M. Hutchins, 17. Chicago: Encyclopaedia Britannica, 1952.

Pratt, H.S. *Manual of the Common Invertebrate Animals.* Philadelphia: Blakiston, 1935.

Rey, Abel. "Vers le positivisme absolu." *Revue philosophique de la France et de l'étranger* 67 (1909): 461–479.

Roger Bacon. *Communium naturalium Fratris Rogeri, partes prima et secunda.* In *Opera hactenus inedita Rogeri Baconi,* ed. Robert Steele, fasc. 2. Oxford: Clarendon Press, 1909.

Ronsard, Pierre de. *Oeuvres complètes.* Ed. Gustave Cohen. 2 vols. Bibliothèque de la Pléiade 45–46. Paris: Gallimard, 1950.

Ruse, Michael. "Biological Species: Natural Kinds, Individuals, or What?" *British Journal for the Philosophy of Science* 38 (1987): 225–242.

Shook, Laurence K. *Etienne Gilson.* Etienne Gilson Series 6. Toronto: Pontifical Institute of Mediaeval Studies, 1984.

Spinoza, Benedict de. *The Ethics [Ethica ordine geometrico demonstrata].* In *The Chief Works of Benedict de Spinoza,* trans. R.H.M. Elwes, 2: 43–271. 2 vols. New York: Dover, 1951. [Reprint of the London: Bell, 1883–1884 edition.]

Thomas Aquinas. *Basic Writings of Saint Thomas Aquinas.* Ed. Anton C. Pegis. 2 vols. New York: Random House, 1945.

—. *De potentia.* Ed. Paul Pession. In *Quaestiones disputatae,* ed. Raimondo Spiazzi, OP, 2: 1–276. 8th rev. ed. 2 vols. Turin: Marietti, 1949.

—. *De veritate.* Vol. 1 of *Quaestiones disputatae,* ed. Raimondo Spiazzi, OP. 8th rev. ed. 2 vols. Turin: Marietti, 1949.

—. *In Aristotelis libros Peri hermeneias et Posteriorum analyticorum expositio.* Ed. Raimondo Spiazzi, OP. Turin: Marietti, 1955.

—. *In Aristotelis librum De anima commentarium.* Ed. Angelo-Maria Pirotta, OP. 4th ed. Turin: Marietti, 1959.

—. *In decem libros Ethicorum Aristotelis ad Nicomachum expositio.* Ed. Raimondo Spiazzi, OP. Turin: Marietti, 1949.

—. *In Duodecim libros Metaphysicorum Aristotelis expositio.* Ed. M.R. Cathala, OP, and Raimondo Spiazzi, OP. Turin: Marietti, 1950, 1964.

—. *In octo libros De physico auditu sive physicorum Aristotelis commentaria.* Ed. Angelo-Maria Pirotta, OP. Naples: M. d'Auria Pontificius, 1953.

—. [*Summa contra Gentiles*] *Liber de veritate Catholicae fidei contra errores infidelium, qui dicitur Summa contra gentiles.* Ed. Peter Marc, Ceslao Pera, and Peter Caramello. 3 vols. Turin: Marietti, 1961–1967. [Trans. by Anton C. Pegis, J.F. Anderson, Vernon J. Bourke, and Charles J. O'Neill as *On the Truth of the Catholic Faith: Summa contra gentiles,* 4 vols. in 5 (Garden City, N.Y.: Hanover House, 1955–1957); repr. under the title *Summa contra gentiles,* 4 vols. in 5 (Notre Dame, Ind.: Notre Dame University Press, 1975).]

—. *Super librum De causis expositio.* Ed. H.D. Saffrey, OP. Textus philosophici Friburgenses 4–5. Fribourg: Société philosophique; Louvain: Nauwelaerts, 1954.

—. *Summa theologiae.* 5 vols. Ottawa: Dominican College, 1941–1945.

—. *Tractatus de spiritualibus creaturis: Editio critica*. Ed. Leo W. Keeler. Pontificia universita gregoriana: Textus et documenta: Series philosophica 13. Rome: Pontificia universita gregoriana, 1937, repr. 1959.

William of Ockham. *Scriptum in librum primum Sententiarum ordinatio* [*Sent.*]. Ed. Gedeon Gál, Stephen F. Brown, Girard J. Etzkorn, and Francis E. Kelley. 4 vols. *Opera Theologica*, ed. Gedeon Gál et al., 1–4. St Bonaventure, N.Y.: Franciscan Press, 1967–1979.

William of Tocco. *Ystoria sancti Thome de Aquino de Guillaume de Tocco (1323)*. Ed. Claire le Brun-Gouanvic. Studies and Texts 127. Toronto: Pontifical Institute of Mediaeval Studies, 1996.

Wolff, Christian. *Philosophia prima sive ontologia*. Ed. Joannes Ecole. Christian Wolff: Gesammelte Werke 2, Lateinische Schriften 3. Hildesheim: Olms, 1962. [Reprint of the 2nd ed., Frankfurt; Leipzig: Officina Libraria Rengeriana, 1736.]

Virgil. *Eclogues*. In *Virgil*, ed. with trans. H. Rushton Fairclough, 1: 1–77. 2 vols. Loeb Classical Library. New York: Putnam; London: Heinemann, 1916–1918; repr. 1922–1925.

Xenophon. *Memorabilia* [*The Memorable Things of Socrates*]. Ed. Josiah Rennick Smith. Greek Texts and Commentaries. Leipzig: Teubner, 1889.

Index